Lizards

Lizards
Volume 2

**Monitors, Skinks, and Other Lizards
Including Tuataras and Crocodilians**

Manfred Rogner

Translated from the
original German by
JOHN HACKWORTH

KRIEGER PUBLISHING COMPANY
MALABAR, FLORIDA
1997

Cover illustrations:
Front: *Varanus giganteus* (photo Eidenmüller)
Back left: *Mabuya quinquetaeniata* (photo Sauer)
Back right: *Varanus griseus* (photo Sauer)

Original German Edition, *Echsen II*, 1994
First English Edition 1997
Printed and Published by
**KRIEGER PUBLISHING COMPANY
KRIEGER DRIVE
MALABAR, FLORIDA 32950**
Copyright © German Edition 1994 by Eugen Ulmer GmbH & Co.
Copyright © English Edition 1997 by Krieger Publishing Company

All rights reserved. No part of this book may be reproduced in any form or by any means, electronic or mechanical, including information storage and retrieval systems without permission in writing from the publisher.
No liability is assumed with respect to the use of the information contained herein.
Printed in the United States of America.

FROM A DECLARATION OF PRINCIPLES JOINTLY ADOPTED BY A COMMITTEE OF THE AMERICAN BAR ASSOCIATION AND COMMITTEE OF PUBLISHERS:

This publication is designed to provide accurate and authoritative information in regard to the subject matter covered. It is sold with the understanding that the publisher is not engaged in rendering legal, accounting, or other professional service. If legal advice or other expert assistance is required, the services of a competent professional person should be sought.

Library of Congress Cataloging-In-Publication Data

Rogner, Manfred.
   [Echsen. English]
   Lizards / Manfred Rogner ; translated from the original German by John Hackworth. — 1st English ed.
      p.   cm.
   Includes bibliographical references (p.   ) and index.
   Contents: v. 1. Husbandry and reproduction in the vivarium. Geckoes, flap-footed lizards, agamas, chameleons, and iguanas. — v. 2. Monitors, skinks and other lizards, including tuataras and crocodilians.
   ISBN 0-89464-972-8 (set : cloth : alk. paper). — ISBN 0-89464-939-6 (v. 1 : cloth : alk. paper). — ISBN 0-89464-968-X (v. 2 : cloth : alk. paper)
   1. Lizards as pets. 2. Lizards. I. Title.
SF459.L5R6313   1997
639.3′95—dc20   95-31852
                                                                                     CIP

10 9 8 7 6 5 4 3 2

Contents

Preface ... vii
Translator's Note ... ix

Lizards 2 ... 1
Knob-Scaled Lizards ... 1
Beaded Lizards ... 3
Monitor Lizards ... 9
Alligator Lizards ... 37
Alligator Lizards and Glass Lizards ... 38
Slow Worm ... 43
Plated Lizards ... 45
Cordylids ... 66
Lacertids ... 82
Skinks, Smooth Lizards ... 169
Giant Skinks ... 170
Skinks ... 188
Slender Skinks ... 202
Night Lizards ... 224
Teiids ... 229
Worm Lizards ... 243
True Worm Lizards ... 244
Florida Worm Lizards ... 247
Sharp-Tailed Worm Lizards ... 247

Tuataras ... 249

Crocodilians ... 253
Alligators and Caimans ... 255
Crocodiles ... 268
Gharials ... 281

Bibliography ... 285
Photographic Sources ... 300
Index ... 301

Preface

This book deals on the one hand with the lizard families that were not considered in Volume 1 and, at the request of my publishers, with the Amphisbaenids, Tuataras, and Crocodilians.

In this work I have endeavoured to assemble all previous and current herpetocultural experiences. Unfortunately, however, only a very small number of herpetoculturists have published the results of their breeding successes. Amongst the lizards that were previously kept in vivaria there were a number of strictly protected species; this is certainly true of the Tuataras and Crocodilians. The sizes of the vivaria mentioned in this work are in no way definitive, since for many years a number of species have been successfully kept and bred under conditions other than those described here.

Some readers may be rather surprised that in this work there is no mention of the various conservation legislation and legal requirements involved when certain species are bought or are obtained in other ways. These regulations and requirements have been the subject of great discussion for some time. Rather than list the legal requirements of each individual species, it is much better to enquire of the relevant authorities before a species is obtained. It is also advisable for prospective purchasers to familiarise themselves with the legal requirements currently in force.

Once again I have to thank a number of people who gave me considerable support and assistance during the compilation of this volume. First and foremost, I must thank my wife, Heidi. Yet again she undertook a number of translations and relieved me of many peripheral problems.

In addition, I must thank Dr. Wolfgang Böhme, Ph.D. (Bonn); Wolfgang Bischoff (Bonn); Dr. Joachim Broch (Stuttgart); Bert Eidenmüller (Frankfurt); Siegfried Ebbert (Bruch); Michael Franzen (Bonn); Klaus Hirschfeld (Oberkirch); Harald Jes (Cologne); Bert Langerwerf (Alabama, U.S.A.); and Dr. Helmut Mägdefrau for their willing cooperation.

Amongst the many illustrations there are many that are published for the first time. For this I have to express sincere thanks to the photographers concerned; they are listed elsewhere in this volume.

Special thanks are also due to my German publishers Verlag Eugen Ulmer (Stuttgart) and to Dr. Heinz Wermuth (Freiberg) for initiating this project.

Translator's Note

I am honoured and delighted to have been commissioned by Krieger Publishing Company to translate Volume 2 of Manfred Rogner's scholarly work *Lizards* and trust that I have done full justice to this informative and masterly work.

Although some species have a common name in German, this is not necessarily the case in English and vice versa. In some cases there is no common name in either language.

Once again, my niece, Nicola Wilkinson, B.Sc., came to my assistance with much biological material, as did Andrew Young, B.Sc. of Glasgow, Scotland, by providing much geographical and meteorological information. Yet again I was willingly and ably assisted by my friend Andrej Koralewski of Lünen, Germany, a gifted linguist and translator in his own right, as is Olaf Kannchen of Hamm, Germany, who also provided answers to several complicated grammatical problems.

David Elks provided material assistance for which I am greatly obliged, and Ian Burn of Newcastle upon Tyne was brave enough to attempt to decipher my hieroglyphic handwriting to type the manuscript. I am grateful to both for their generosity and patience.

As always, my friend Peeter Põldsam, the foremost Estonian herpetologist and reptile keeper, made many helpful suggestions for which I am particularly indebted.

Without the cooperation of those mentioned above, this project, like its companion volume, would have taken much longer to reach completion.

John Hackworth
Newcastle upon Tyne, England
1996

Lizards 2

KNOB-SCALED LIZARDS
Family Xenosauridae

Fossils of *Xenosauridae* have even been found in the upper chalk layers of Europe. Nowadays, relicts of this family are only known from Mexico, Guatemala, and China. These belong to two genera with only four recent species. Very little is known about their lifestyle in the wild.

Shinisaurus Ahl, 1930
Crocodile Lizards

The sole area of distribution for this monotypic genus is in southern China. Here, this very rare lizard lives in moist forests in the vicinity of water. The skull is relatively high and narrow. The double row of protruberant scales along the tail is somewhat reminiscent of a crocodile, thus giving these lizards their common name.

Shinisaurus crocodilurus Ahl, 1930
Chinese Crocodile Lizard

Distribution: Only a very small area of the Guanxi Province of southern China (Mägdefrau, 1987).
Description: According to Mägdefrau (verbal communication), this lizard may reach a total length of up to 41.5 cm. Its colouring is extremely variable. The upper side of the animal is almost always a uniform brown, whilst the underside may be yellow, orange, or red, sometimes with a grey throat. A spotted underside is not uncommon. The males are usually more brightly coloured and generally have a reddish tone. There are two large black spots in the shoulder region.
Habitat: Along slow-flowing streams or beside ponds with densely overgrown banks, at altitudes of 500–700 m above sea level. In these places the temperatures in summer may rise to 40 °C, whilst in winter they may fall to the freezing point. In this region the summer is very wet and the autumn very dry, but this does not have any significant effect on the waters.
Lifestyle, husbandry, and reproduction: This species is strongly tied to water and will often spend long periods lying motionless in this element. Otherwise they may

Shinisaurus crocodilurus

be found on rocks or in trees in the vicinity of water. In such places they may appear to be sleeping, but at the slightest hint of danger they will leap into the water. They will inflict a powerful and painful bite without any warning. This species is also an excellent and elegant swimmer which propels itself through the water with side-to-side motions of the longitudinally compressed tail, whilst its legs lie neatly alongside the body.

In the wild this lizard feeds on fresh-water crayfish, small fish, young frogs, tadpoles, earthworms, insects, and young mice. They will often lie with closed eyes, totally submerged in water for up to 30 minutes. They even slough completely in water. When the temperature drops to around 9 °C in late autumn, they begin their 3- to 4-month hibernation. In Mägdefrau's case, hibernation began at 12 °C.

For this species a roomy aquavivarium with a large water section and several stout climbing branches is required. Favoured places will soon be identified and sought out daily. A built-in waterfall with bathing pool has also proved to be expedient. Additional furnishings are sturdy plants, rocks, and branches because this species enjoys climbing. The prefered basking place should be heated by a HQL lamp.

Even outside the breeding season this is a very aggressive species and the males are especially intolerant of one another. Wick (verbal communication) feeds his lizards on beetles, locusts, 1- to 2-week-old mice, crickets, and earthworms. He keeps two females and one male in an aquavivarium, the water section of which covers three-quarters of the total ground area. A 4-month hibernation at 12 °C is vital; otherwise the females will not be willing to mate. The males, on the contrary, appear to be willing to mate throughout the year. As far as is known, gravidity lasts 8 to 9 months. In Wick's case, the first clutch produced six young and the second, seven. The young weighed around 4 g and measured between 12.5 and 13.3 cm. Females are quite capable of producing clutches of up to 15 young.

In Mägdefrau's case the smallest of the young measured 10.5 cm and the largest 15 cm. The heaviest of the young weighed 5.6 g. Rearing the young is not without problems because they will not always accept insect larvae, small crickets, or waxmoth larvae and must sometimes be force-fed. After around 6 weeks the males should be removed and reared separately; otherwise conflicts will inevitably occur. Otherwise the young should be kept in the same way as the adults.

BEADED LIZARDS
Family Helodermatidae

Of the Helodermatidae, there remains only one genus with two species. Beaded Lizards are known as fossils from the Oligocene periods of North America and Europe. They are the only remaining venomous lizards.

Heloderma Wiegmann, 1829
Beaded Lizards

Both species of Beaded Lizards belong to the genus *Heloderma*. They live in the southwest of the United States and in Mexico. Beaded Lizards have a solid, stout body and a broad, flat head with a blunt, rounded snout. The limbs are short and powerful with strong claws. The rounded tail is very plump and serves for fat storage. They have only very small eyes and slitlike ear openings. Beaded Lizards have a fleshy, deeply forked tongue. The body is covered in semi spherical scales that are not arranged in a particular pattern.

Heloderma horridum (Wiegmann, 1829)
Beaded Lizard

Distribution: Western Mexico.
Description: The Beaded Lizard reaches a total length of up to 80 cm. On a dark

Heloderma horridum

brown or black background there is an irregular number of variously sized, more-or-less transversely arranged, pale yellow, but sometimes red, blotches. The blotches are less pronounced than in *Heloderma suspectum*. On the tail the blotches form four to seven transverse bands. The tail is as long as the body or even a little longer. Males have slightly stronger limbs, a broader skull, a longer neck, and a somewhat more sturdy body than females (Trutnau, 1976).

Habitat: Dry, light forests.

Lifestyle, husbandry, and reproduction: The prefered temperatures of the Beaded Lizard are 26 to 28 °C. In this range, the lizards are active during the day as well as at night and will frequently burrow into the sand. For further information see Gila Monster *Heloderma suspectum*.

In accordance with its size, this species requires a large, dry vivarium. A suitable substrate is a layer of sand and gravel, some 10 to 20 cm deep. The furnishings consist of a water bowl and some flat, stacked rocks firmly cemented together. The ground-heating cable should also be firmly cemented-in. Because of the conditions prevalent in their natural habitat, bright lighting is also essential. During the day, the temperature should be 25 to 30 °C, falling to around 18 °C at night. During the winter period, these values should be reduced to 20 °C and 12 °C.

The activity period stretches from spring to autumn, during which time the lighting and ground heating should be in operation for 10 to 14 hours daily. Trutnau's lizards of this species are fed on diced beef heart, eggs, hamsters, young guinea pigs, young rabbits, and pieces of rabbit meat. However, their main diet is mice and young rats. Before being given, the food is dusted with a polyvitamin preparation and is given in a 10- to 14-day cycle when each animal is given two to six mice.

Seasonal temperature variations are not only important for breeding this species, they also bring the lizards into better condition. In Trutnau's case, after hibernation, the male followed the female relentlessly, inflicting a bloody wound on the neck. Mating was observed during the evening of June 20. On August 11 1970 the female laid four eggs, three in the morning and one in the afternoon. On the following day a further egg was produced. The oval parchment-like eggs measured between 5.9 and 6.4 cm and lay uncovered on the substrate. Unfortunately, it was not possible to incubate the eggs successfully.

It was only on 15th June 1974 that a clutch of ten eggs was secured and subsequently successfully incubated. The temperatures varied between 20 and 30 °C but were usually around 25 to 27 °C. On 12th November one young Beaded Lizard peeked from a small slit at the centre of the egg. Unfortunately Trutnau severed the umbilical cord and the young died three days later measuring 15.8 cm and weighing 32 g. Previously six infertile eggs had to be removed after two or three months and the young in the remaining three eggs suffocated.

Heloderma suspectum Cope, 1869
Gila Monster

Distribution: Southwestern United States, northern Mexico.
Description: Essentially the same as *Heloderma horridum*, but this lizard only reaches a total length of 60 cm. The tail is only around two-thirds of the snout-vent-length (SVL), and the elongated flecks are an orange-yellow to flesh colour.
Habitat: Desert regions.
Lifestyle, husbandry, and reproduction: The short activity period of the Gila Monster is limited to the breeding season in spring and the rainy season of July and August. The remainder of the year is spent in burrows which they dig themselves. Here they live off the fat reserves in the tail. The eggs of other reptiles and ground-nesting birds form the main part of the diet, but these lizards will also eat young mammals and birds. The Gila Monster does not eat carrion. Once the lizard seizes its prey it does not release it. This allows the venom to flow through two 40 × 5 mm venom glands into the victim. This process also serves as its first line of defence. The venom glands lie at the rear of the lower jaw. From here the venom flows, as a result

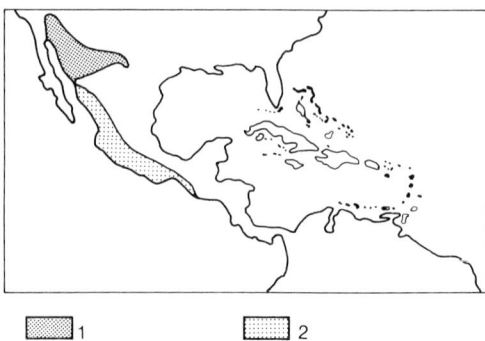

Distribution ranges of *Heloderma suspectum (1) and Heloderma horridum (2)*

of contractions of the jaw muscles, into a channel between the jaw and lips to the teeth which are curved slightly backwards. The teeth always have a longitudinal groove at the front and usually also at the back. Some of the teeth in the upper jaw are also grooved and are well anchored into the gums so that when the animal bites the teeth are pushed backwards. It is only then that the teeth may penetrate up to one centimetre deep into the flesh of its victim or an aggressor. In one bite, these lizards are able to deliver up to 20 drops of venom. The mixture of venom and saliva flows through the "tooth grooves" into the wound. Only around 0.005 mg of dried venom from this species is fatal for a human. The venom works essentially on the central nervous system and by causing paralysis of the breathing system can lead to eventual death. The effects of this venom in combination with alcohol should never be underestimated.

Although there are very few recorded cases of bites causing human fatalities, the symptoms of a bite from this species can be extremely unpleasant. Phisalix (quoted in Klingelhöfer, 1957) was bitten by a *Heloderma suspectum* whereby the bite was so firm that a tool had to be used to lever the jaws apart. Soon afterwards, unbearable pain started, the fingers and part of the hand became swollen and inflamed whilst the actual position of the bite was pale. Several fainting fits also occurred. Mitsch (quoted in Klingelhöfer) was bitten 10 cm above the wrist. The teeth penetrated to a depth of some 8 mm. After 20 minutes, the victim was in considerable pain, his arm was swollen and after 30 minutes severe anxiety set in. Profuse sweating, reduced inhalation, and prolonged losses of consciousness were further symptoms of envenomation. The heart rate slowed down, headaches were frequent with increasing sensitiveness, and for 24 hours the victim was unable to urinate. The arm in question remained swollen for more than 1 week.

It is therefore for good reason that this otherwise peaceful lizard shows obvious warning colours.

On May 20, 1963, three *Heloderma suspectum* arrived at San Diego Zoo from

Tucson, Arizona. Two lizards were observed mating only 6 days later during the afternoon of May 26. On July 25, the female laid five immaculately white eggs. It is not known however whether these eggs were fertilised on May 26 in captivity, or earlier in the wild. The eggs measured 61.1 to 66.8 mm in length and 32.0 to 32.6 mm in width and weighed 37.2 to 40.8 g. The eggs were transfered to two sand-filled boxes and incubated in the increased humidity of the corridor of the reptile house. There, the temperature was a minimum of 27 °C and sometimes reached a maximum of 32 °C. During the weekly inspection of the eggs, the upper layer of sand was swept away using a fine brush. If the humidity was too low, this was indicated by longitudinal depressions in the shells. One egg collapsed, but when opened for inspection proved to be fertile.

After 92 days the eggs were again measured and weighed. They had increased by 0.5% in length, 4.6% in width and 6.04% in weight. The first egg containing a live young hatched after 117 days.

On November 20 one egg was opened and found to contain a dead young, the head of which, as in the case of the first hatchling, was on the opposite side of the egg. The next egg to be opened, and from which the head of a young was protruding, was found on November 25, i.e., on the 124th day at 0700 h. At noon the following day this young also lay in a similar position in the egg. The opening in the egg was made larger after which the young lizard was able to free itself. The penultimate young hatched on November 27 (126th day), and the last 4 days later. The young had total lengths of 150 to 164 mm. The tail accounted for 42 to 49 mm and the young weighed from 28.1 to 32.9 g. On December 29, the three young were fed for the first time. Each received a dead mouse. This did arouse interest, resulting in tongue flicking, but the mice were not eaten. A live pink mouse dipped in egg yolk was offered to each of the young the following day. This was immediately recognised as food and eaten. The young subsequently ate mice that had not been dipped in egg yolk.

Regarding the incubation time and the time it takes for the young of this species to actually hatch, it would appear that, at least with this species of Beaded Lizard, in the wild the young would either remain in the egg until the following spring or hibernate as hatchlings. This phenomenon is not unusual for reptiles in the temperate regions of the world.

In connection with this prolonged hatching process, the director of the Arizona Sonora Desert Museum, William Woodin, remarked that he had never seen newly hatched Gila Monsters in the wild in autumn. All young given to the museum had been found in spring or summer, i.e., between May and August. It is thus most likely that the eggs themselves hibernate since it is most unlikely that the young would hatch during the coldest part of the year. Otherwise young specimens would certainly have been found on the first warm days of the year (Shaw, 1968).

Gila Monsters should be kept in the same way as the previous species. A water container is important and its contents should be kept at a temperature of at least 25 °C. These animals are usually active at dusk or during the night. In spring they are also active during the day. From November to February they require a winter rest at lower temperatures. If the atmospheric humidity is too low, Gila Monsters will lie in the water container. In Trutnau's case (1968), his Gila Monsters were given mainly eggs, birds, and nestling rodents. The eggs were licked out after they had been bitten and cracked against a solid and uneven surface. The contents of the eggs were allowed to flow into the stomach by raising the head. Rodents were bitten and held fast until the venomous saliva took effect. This may last from 5 to 30 minutes. Rodents are usually swallowed head-first. Only very rarely are they swallowed rump-first.

Eidenmüller & Wicker (1992) also report on the breeding of one young of this species. The authors keep their lizards in a vivarium measuring 150 × 60 × 60 cm. Sand forms the substrate whilst a tree root and several large pieces of cork bark provide hiding places. The provision of a water container measuring 40 × 20 × 15 cm also proved to be expedient. This is regularly sought out by the animals which often spend several days lying in it, submerged up to their nostrils. In the wild this behaviour is not possible. Thus in the Exotarium of Frankfurt Zoo, they were given two artificial caves containing moist substrate. The caves have removable covers so that additional water may be added to the substrate when necessary. Lighting is provided by a mercury vapour lamp (Philips HPL, 50 W™) and temperatures vary between 22 and 25 °C. A spotlight (Osram Concentra™) serves as an additional heat source for basking. During the summer months the lighting is in operation for 14 hours, whilst during spring and autumn this period is reduced to 8 hours. Each lizard spends the winter individually at 13 to 15 °C in a Macrolon™ box containing dry wood shavings and a small water bowl.

After hibernation, the male was moved to a vivarium measuring 140 × 45 × 40 cm. In mid-April, a semi-adult female was introduced. Only a few minutes later, the male left his refuge and, with its tongue flicking wildly, proceeded to follow the female who soon gave up all attempts to escape. After 3 hours the pair were seen mating for several hours. A further copulation was observed on May 3, and on June 15 two soft-shelled eggs were found beneath a piece of cork bark around midday. Each egg measured 6.5 × 3.0 cm. The eggs of Helodermatids are very sensitive to contact with moisture. These two eggs, and a further egg laid later, were placed in an incubation box containing only very slightly moist Perlite™. The incubator was set at 27.5 °C and an atmospheric humidity of 90%. Two of the eggs rotted and had to be removed. After 4 weeks the temperature was increased to 29 °C and on October 21, 1990, the head of the young could be seen protruding from the shell. Two days later, i.e., after 130 days, the young left the egg measuring 16.8 cm and weighing 38.2 g. From the fourth day after hatching, it willingly ate vitaminised nestling mice. At the age of around 6 months, on April 15, 1991, it had attained a length of 22.1 cm and weighed 71.5 g.

Helodermatids have been kept in captivity for up to 20 years (Trutnau, 1984).

MONITOR LIZARDS
Family Varanidae

Monitor Lizards live in Africa, Southern Africa, Asia, and Australia. Their skull is completely ossified and thus protected from below when they are eating large pieces of food. Their throat can be enlarged by dislocation of the thyroid bones. Monitors swallow their food whole or in large pieces. The shape of the vertebrae, the chambering of the heart, and the absence of a bladder are characteristics that prove that Monitor Lizards and giant snakes are related. The family Varanidae contains only one recent genus: *Varanus*.

Varanus Merrem, 1820

The approximately 30 species of this genus are in some ways so different or have such great similarities that we nowadays divide them into 10 subgenera. Even today, Monitor Lizards may occasionally present surprises. Whilst making a film about northern Yemen, a sequence containing a Monitor Lizard was also coincidentally filmed. Subsequent examination showed that this creature was a previously unknown species, the Yemen Monitor (Böhme, Fritz & Schütte, 1987).

Monitors come in sizes ranging from 20 cm to 3 m and with weights ranging from 20 g to 150 kg. They also live in the most divergent habitats. Despite the great difference in sizes, monitors are generally strongly built creatures with strong, five-toed limbs ending in sharp claws. The large pointed head is usually held high on an elongated neck. Round pupils and a well-developed ear opening are common to all species. The nostrils of some species however may be quite different. The strong teeth are curved backwards slightly whilst the tongue is long and deeply forked. Except in the case of *V. brevicauda*, the tail is usually longer than the SVL. In some species the tail length may even be double the SVL as is the case with *V. salvator, V. prasinus, V. a. acanthurus, V. glauteri* and *V. glebopalma*. In cross section the tail may appear to be round or somewhat laterally compressed, in which case the upper side is keeled. The small scales on the body are not arranged like roof tiles and do not overlap. Some species, however, do have scales that are longitudinally keeled. Some enlarged tubercular scales may be found in the neck region of some species.

Some species lead a more-or-less aquatic lifestyle, whilst others live in deserts and on dry steppes. Still others live in light forests and on savannahs. In amphibious species, the tail is laterally compressed and serves as a paddle when swimming. The Nile Monitor may submerge for up to 1 hour.

According to their size, diurnal monitors feed on anything from insects to medium-sized vertebrates. They will however also devour eggs and carrion. *Varanus dumerilii*, a feeding specialist, eats mainly crayfish. *Varanus olivaceus* will also eat fruit.

When in competition, males will rear up, shoulder to shoulder, each attempting to throw the other to the ground. Only rarely however are wounds inflicted in these conflicts.

In the wild, monitors reach sexual maturity at 3 to 5 years old. In captivity however, this frequently occurs earlier. When mating, males use their hindlegs to hold the female in position, whilst simultaneously licking her in the neck region. Only very rarely do they bite the neck of the female, a phenomenon which is quite common amongst all other lizards. According to their size and age, female monitors may lay clutches varying from 7 to 60 eggs. Some species bury their eggs, whilst others conceal them in hollow trees. Still other species deposit their eggs in termite hills.

Many species of Monitor Lizards have not only been successfully kept in captivity for a number of years, but it has also been possible to breed a number of species under captive conditions.

Varanus acanthurus Boulenger, 1885
Ridgetail Monitor

Distribution: Northwestern Australia, through the Northern Territories, the Mount Isa district of Queensland including the Monte Bello Islands and Groote Eylandt in the Gulf of Carpentaria.
Description: The brown to reddish brown Ridgetail Monitor reaches a length of over 60 cm. The head and neck have yellowish longitudinal stripes; along the back these change to vivid eyespots. The tail has transverse bands and the white underside may have pale spots.
Habitat: Dry areas, open bushland, and stony semi-desert regions.
Lifestyle, husbandry, and reproduction: This lively and active lizard enjoys burrowing and therefore requires a large territory. Because of their small size, they are a very popular species for keeping in captivity. They are furthermore quite peaceful, making it possible to keep several specimens together in one vivarium.

This species of Monitor Lizard requires a spacious vivarium with a 1:10 mixture of sand and loam as substrate. This species digs tunnels in the substrate and will use these as refuges. A water bowl containing fresh water should always be available. The water should be changed daily. During the day, temperatures of 26 to 30 °C are necessary, whilst beneath a spotlight it should be even warmer. At night the temperature may be allowed to drop to 22 °C. Food in the form of crickets, other insects, nestling mice, and some egg should be given. All food should be coated with a multivitamin preparation. This species will also devour addled pigeon eggs, without the shell, and diced beef heart (Erdfelder, 1984).

Although at the Wilhelma Zoo in Stuttgart (Germany) the eggs were left to hatch in the vivarium, the keeper should always attempt to incubate the eggs artificially.

Lizards

Varanus acanthurus

Around 4 weeks after mating, the girth of the female increases considerably, and the outline of the eggs may be clearly seen through the wall of the stomach. At this point, the female should be transferred to an empty vivarium with ground heating and a 25 cm layer of sharp, slightly moist sand as substrate. In Erdfelder's case, a female laid eight white eggs at a depth of 15 cm. These measured around 32 to 33 mm in length, 18 to 19 mm in width, and weighed around 6 g. They were incubated at 29 °C and an atmospheric humidity of 95% in a moist mixture of sharp sand and peat in which they were buried 5 cm deep. After 140 days the head of the first young was seen protruding from the egg. It left the egg the following day, measuring a total length of 172 mm. Within 4 weeks, seven young hatched. One egg was infertile. The young measured from 145 to 172 mm and weighed between 7 and 9.5 g. The long periods between individual hatchings were probably caused by temperature variations in the substrate. After the second day the young fed voluntarily and were also given a shallow water bowl from which they regularly drank. After around 11 months, they had grown to between 29 and 42 cm and weighed 28 to 78 g. At around 15 months of age, the largest of the young was already sexually mature. In the summer of 1983, the largest of the captive-bred females laid four eggs from which three young hatched after 138 days.

originated from a further clutch that had been laid unnoticed, around 14 days earlier in another position in the vivarium. Schmida estimates that the incubation period lasts from 10 to 12 weeks. The young voluntarily ate small moths and newly moulted grasshoppers. They had a total length of 80 mm, of which 38 mm was tail. In a further report the author writes that during the period of mid-September until the end of October, a female laid a total of eight eggs in three clutches, from which young hatched after approximately 10 weeks. The young measured on average 77 mm (Schmida, 1974, 1975).

Varanus caudolineatus Boulenger, 1885
Stripe-Tailed Monitor

Distribution: Western Australia.
Description: This small species only reaches a length of around 30 cm and has a pattern of irregularly arranged dark brown spots. The upperside is light grey to beige whilst the underside is cream with a feint pattern of dark flecks, especially on the throat. Males appear more robust than females. In addition they have longer hind legs and a longer tail (Schmida, 1975).
Habitat: Semi-desert regions.
Lifestyle, husbandry, and reproduction: This species is often found on or in hedges of *Acacia* or on dead *Casuarina*. Sometimes they may even be found in pairs in a hole. They are a joy to keep and are quite happy in a vivarium with temperatures between 20 and 25 °C. Several climbing branches, if possible hollow to provide refuges, should be provided. This monitor prefers refuges that are only barely large enough for it. They will eat all manner of insects as well as meat cut into fine strips. This will be taken from forceps. In Schmida's case, a semi-adult female laid a clutch of three infertile eggs measuring 22 × 10 mm. Nothing more is known to date about the breeding of this species.

Varanus dumerilii (Schlegel, 1839)
Dumeril's Monitor

Distribution: Malacca Peninsula, Sumatra, Mergui Archipelago, Rhio Archipelago, Batu Islands, Banka, Biliton, and Kalimantan (Borneo).
Description: This monitor reaches an average length of 130 cm with the maximum being 150 cm. It is impossible to distinguish the sexes externally. The brownish coloured animals have four or five narrow yellowish bands and a light forked marking on the neck. The neck region has large strong scales. On a short, flat skull, the nostrils are slit-shaped. The strongly compressed tail is longer than the SVL. The young appear black and have well-defined transverse bands. The top of the head is usually yellow to vermillion in colour.
Habitat: River banks and rain forests with large mangrove areas.

Lizards

Varanus dumerilii

Lifestyle and husbandry: For some time it was thought that this lizard was a diurnal tree dweller but actually very little was known about its life in the wild. It was thought that they ate insects, birds, and eggs in the wild and would not eat meat and mammals in captivity. However, further examination of special characteristics of the skull, the body shape, and the natural habitat of this species proved that Dumeril's Monitor is an amphibious tree dweller which appears to feed on crabs (Krebs, 1979). Rese (1984a) describes how the reluctance of a Dumeril's Monitor to feed was finally broken by offering crabs. If crabs are given, the animals may also be fooled into eating mice by scenting these food items with crab meat. Because of their size, these creatures require spacious vivaria with a large water container and climbing branches. Should any of these animals ever be exported from their natural habitat, they should only be kept by experienced monitor enthusiasts.

Varanus exanthematicus Bosc, 1792
Bosc Monitor, Savannah Monitor

Distribution: West and central Africa, southwards to Zaire and northern Angola. Several subspecies are known. The nominate form may be found from Senegal to Ethiopia, *V. e. albigularis* from South Africa to northern Ethiopia. As further subspecies, *V. e. microstictus* and *V. e. angolensis* have variations in scalation and *V. e. ionides* has a different juvenile pattern.
Description: *V. e. albigularis* reaches a length of 150 to 200 cm, whilst the nominate form is only a little over 100 cm. It is suspected that at a length of 60 cm these creatures are sexually mature. The nominate form has a more uniform colour and pattern. The white throat is absent and the scales on the neck are enlarged. *V. e. albigularis* shows greater contrasts, with a light belly, a lighter throat, and smaller scales in the neck region.
Habitat: Dry areas, especially steppes and deserts.
Lifestyle, husbandry, and reproduction: The Bosc Monitor avoids water and overgrown areas (Rese, 1983b). Because of their size, these lizards require a very spacious vivarium with temperatures between 28 and 35 °C. Duinen (1983) housed his animals superbly in a vivarium with a ground area of 10×5 m, as did Koore (1988), using an area of $500 \times 250 \times 280$ cm. Sand is a suitable substrate. Several tree trunks and large branches lying on the substrate provide refuges for the lizards. In the wild, these monitors feed on insects, beetles, ground-dwelling birds, reptiles, toads, eggs, and carrion.

Males claim a territory that they will defend vigourously. If two males encounter one another they will make mutual threats, each attempting to intimidate the other. If this fails, there follows a vicious fight, during which the two bodies will be intertwined and severe bites inflicted. Before mating, the male will follow the female relentlessly, occasionally biting his intended partner on the neck and scratching her back and legs with his strong claws until eventually mating takes place. According to Patterson & Bannister (1987) the female subsequently lays 20 to 50 eggs in a hole that she digs herself. When keeping these lizards in the vivarium, it should be borne in mind that in the wild these monitors are subject to considerable climatic variations.

A male *V. e. albigularis* and a female *V. e. microstictus* inhabited a vivarium of 10 m^2. The vivarium and the container intended for egg laying were sprayed daily. When the keeper saw the female digging in the container of fine gravel, he found a supposedly newly laid clutch of eggs and transfered the egg container to an incubator. The clutch contained 12 eggs buried at a depth of 20 cm. After only a month, a newly hatched young was found in the incubator. Because incubation normally takes 5 to 6 months, it was obvious that the eggs must have lain much longer in the egg container where they had started to develop. The hatchling had a total length of 15 cm with a SVL of 7 cm and a tail of 8 cm. Two other hatchlings were 1 cm longer.

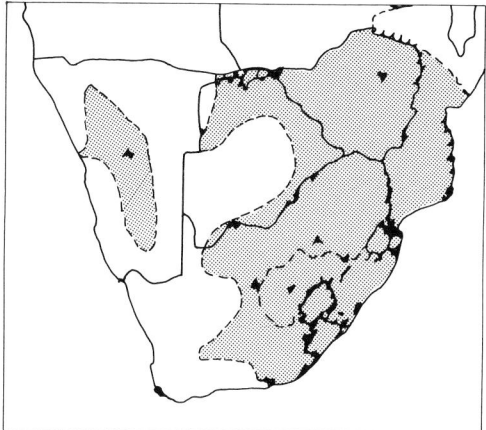

Distribution ranges of *Varanus e. albigularis*

From the third day after hatching the young greedily ate small house crickets from forceps. At 30 °C, development of the eggs usually takes between 146 and 167 days (Duinen, 1983).

Varanus flavescens (Hardwicke & Gray, 1827)
Yellow Monitor

Distribution: From Pakistan in the west to Bengal in the east, whereby Sind in Pakistan is the most westerly limit of distribution. The Yellow Monitor has furthermore been found in the Indian province of Bihar and on Gordon Hill.
Description: A relatively small, thick-set species which reaches a total length of only 1 m. The animals are a dirty-yellowish colour. Subadults have poorly defined and insignificant transverse bands. This monitor has a relatively short and high head as well as a tail that is strongly compressed laterally and has slight double keeling.
Habitat: Obviously the Yellow Monitor prefers a much more moist habitat than had been previously thought. This was observed in Rotterdam Zoo and described by Sights (quoted in Visser, 1985).
Lifestyle, husbandry, and reproduction: As in *Varanus bengalensis*. A pair lived in a vivarium of 600 × 250 × 280 cm in the heated Reptile House of Rotterdam Zoo. On the small side of the vivarium there was an observation window. The concrete floor was covered with a 10 cm layer of sand. Twice weekly the monitors were given young rats; adult mice; day-old chicks; and a mixture of ground beef, eggs, and a mixture of vitamin and mineral salts. As soon as the lights were switched on in the morning, the lizards appeared and were active for around 30 minutes whilst they basked beneath the spotlights. They then lay in the water for several hours and were only active again much later. Several matings were observed between the hours of 0800 and

1100 and on July 21, 1983, eggs were finally laid. The 10 eggs were immediately transferred, in a moist mixture of peat and sand, to an incubator. At an atmospheric humidity of almost 100% and at 30 °C, the eggs deteriorated so badly that they had to be placed in individual small containers. In the bottom of each container, a hole was cut. This hole was only so large that it just prevented the egg from falling through. The small containers were then placed inside a second container, the base of which was covered with water. Unfortunately, this arrangement was also not successful and on December 17, i.e., after 149 days, only dead young embryos were found inside the seven eggs. On December 23 a further egg was opened and also found to contain a dead embryo with vertebral deformations. The remaining two eggs contained live young. One of these was badly deformed and had to be euthanased, but the other had a total length of 145 mm and only a very slight deformation of the spine. However, it only survived for a few days.

Varanus flavirufus Mertens, 1958
Gould's Monitor

This species was earlier known by the name *Varanus gouldii*, but for nomenclatory reasons had to be renamed (Böhme, 1991).
Distribution: Australia to the extreme southeast and in some moist areas of the east coast.
Description: This widely distributed species has various geographic variations that differ in colour, pattern, and size. The colour on the back varies from light yellow to predominantly black with light or dark spots and flecks that form irregular transverse bands. Sometimes the cream or yellow-coloured flecks have a dark spot in the centre and look like eyespots. The legs usually have white or yellowish spots. With the exception of the tip, the tail appears light brown to black with light scales that form narrow rings. The tip of the tail is usually white, cream, or yellow, whilst the underside of the tail is a whitish colour. The small scales on the head are smooth and irregularly arranged. At the centre of the body there are 140 to 220 scales. Beyond its root the tail is laterally compressed and about 1.5 times longer than the SVL. The total length is around 1.60 m.
Habitat: Semi-deserts, only rarely in moist forests.
Lifestyle, husbandry, and reproduction: As ground dwellers, these monitors dig deep burrows or use the burrows of other animals. They do however, often hide beneath rocks or wood piles. When seeking food they cover enormous distances, eating insects, other reptiles, birds, and small mammals. Larger specimens often rear up, using their tail as support, to survey their territory. Because of their compulsion to wander, extremely large and spacious vivaria are vital.

Until 1991 this Monitor was still wrongly named *Varanus gouldii*. It was Böhme (in 1991) who first remarked, however, that this designation applied to a species already described as "new" by Storr in 1980, *Varanus panoptes*, the lectotypus of which

(*Hydrosaurus gouldii* [Gray, 1838]) already existed in a British museum. For this reason, the next available name for this species came into force.

Gould's Monitor has already been bred in captivity several times (Barnett, 1979; Irwin, 1986). Irwin, manager of Queensland Reptile Park, north of Brisbane, observed the mating behaviour of three adult males and two females during the breeding season in an outdoor enclosure of 10.5 × 7.4 m. The brick walls had a 1.1 m steel overhang. The substrate was sand, the height of which was raised at the centre for better drainage. Beneath the raised part, the lizards had dug several tunnels up to 3.5 m long. At this point the average length of the Monitors was 85 cm.

At the middle of February 1984, several matings were seen within 12 days. During this time the male behaved very aggressively towards every other lizard in the enclosure, except his partner. Later, the gravid female dug a further series of exploratory holes, and on December 25 laid seven eggs in a 1 m long tunnel which did not touch any of the rocks or the available water source. The incubation medium was vermiculite mixed with an equal amount of water. The temperature in the incubator was 30 to 32 °C and the atmospheric humidity 100%.

On September 16, i.e., after 265 days, the hatchlings emerged measuring on average 27 cm. They were identical to their parents, but more brightly coloured. They were offered food 8 days after hatching. Because they began fighting, they were placed in two separate containers. At the age of 3 months, the average length of the young had reached 33 cm. They were fed on young mice, strips of poultry meat, and canned dog food. The temperature in their vivaria fluctuated between 28 and 33 °C.

Varanus giganteus (Gray, 1845)
Perentie

Distribution: Australia; from Western Queensland to the coast of Western Australia.
Description: These monitors may reach a length of 250 cm and are thus the largest lizards in Australia. Their upper side is brown. Along the back and on the sides they have large cream or yellow spots or flecks which occasionally form transverse bands along the back and tail. On the head, on the sides of the neck, and on the throat there is a reticulated pattern. The limbs are dark brown with scattered cream or yellow spots. In young animals the colours and patterns are extremely attractive but they quickly fade with increasing age. With the exception of its root, the tail is strongly compressed and on its lower reaches has a double row of heavily keeled scales.
Habitat: Dry, hilly, and rugged landscapes interupted by extensive sandy plains with low bushes and few trees.
Lifestyle, husbandry, and reproduction: This monitor is predominantly a ground-dweller that inhabits deep rock fissures and rabbit burrows between rocks. They cover a large territory searching for insects, birds, and small mammals. The 10 × 15 m outdoor enclosure of the Bredl Reptile Park (Renmark, Australia) consists of a

large sandhill, large piles of rocks, and some enormous tree trunks. This accommodated three males and one female of this large species. They were fed twice weekly on newly killed rabbits at a regular feeding place.

The largest of the males, measuring some 180 cm, mated with the 130 cm female. Of the clutch of eleven eggs, ten were recovered, placed in peat and transfered to an incubator. The eggs each measured around 50 × 78 mm and weighed some 82 g. The temperature in the incubator varied between 30 and 32 °C whilst the atmospheric humidity was constant at 85%. After an average duration of 231.5 days and after an incision had been cut in the eggs, the young emerged. The average total length was 375.3 mm and the average weight 40 g. The young began to eat 9 days after hatching, They were first given nestling mice and then adult mice.

Varanus gilleni (Lucas & Frost, 1895)
Pygmy Mulga Monitor

Distribution: Northern South Australia, central Northern Territories to northwestern Western Australia.
Description: Male Pygmy Mulga Monitors may reach a total length of 38 cm, whilst at 28 cm the females remain considerably smaller. The rounded, compact body has relatively short legs. The sandy reddish ground colour has small brown spots. Along the back there are dark brown to lilac transverse bands that reach down to the first half of the tail. On the lower half of the tail the bands change into stripes. The underside is white.
Habitat: This small monitor usually lives beneath the bark of dead trees, especially on Mulga trees (*Acacia aneura*), Casuarins, and Acacias (Peters, 1974; Schmida, 1975).
Lifestyle, husbandry, and reproduction: In the wild, this species feeds on insects, geckoes, and skinks (Peters 1985). Because of their small size, Pygmy Mulga Monitors are much easier to keep in a vivarium than other larger species. They require a dry vivarium containing a layer of sand some 8 to 10 cm deep. Some parts of the substrate should be kept constantly moist. In the natural habitat of these animals, the temperature sometimes rises to 35 °C (Polleck, 1980). Thus in the vivarium, this species requires temperatures between 25 and 30 °C. Under a spotlight the localised temperature should be even higher. It is quite sufficient to feed this species two or three times weekly. They will accept insects and occasionally pieces of meat from forceps.

In June, Schmida observed the mating of these animals after heavy rainfall. A 22 cm male followed a 27 cm female relentlessly for 2 days. During this time several matings took place. To achieve this the male held the female tightly with all four legs and his snout pressed into the neck. The male then wriggled the rear end of the body beneath that of the female until the cloacal openings lay opposite one another. Each copulation lasted up to 30 minutes. After 3 weeks, the female, which had increased greatly in girth, began to dig incessantly in the sand, finally digging a hole into which it disappeared for an entire day. The female finally re-emerged much slimmer and closed the hole. The first 5 cm of the hole was filled with compacted sand, behind which there was a tunnel some 20 cm long. There then followed another 5 cm deep layer of compact sand, behind which, in the final chamber, three parchment-like eggs were found. The eggs were 30 mm long and 14 mm wide. For 2 months the eggs developed well in an incubator at 28 to 30 °C. Unfortunately, the heating failed and the incubation was unsuccessful.

In the case of Horn (1978), a female some 25.4 cm long laid three eggs which were also placed in an incubator. The female was unable to pass the fourth egg and died from egg binding. This clutch was incubated at an atmospheric humidity of 100% and temperatures of 29 to 30 °C. Two of the eggs shrivelled but the third egg increased in size, and after 88 to 94 days slits appeared in the egg from which the head of a hatchling protruded. On the following day the hatchling left the egg measuring 11 cm and weighing 3.9 g.

Varanus glebopalma Mitchell, 1955
Long-Tailed Rock Monitor

Distribution: Tropical Northern Australia from the Kimberley Region in Western Australia into Western Queensland.
Description: This monitor may reach a length of 90 cm. The black upper side of the body has individual reddish brown scales that form a reticulated pattern on the flanks. On the upper outside of the limbs there are cream or reddish brown spots or flecks on a black background. On the legs these change into large patches. The first half of the tail is predominantly black whilst the second half is a cream to yellow colour. On the white throat there is a wide, reddish brown and light coloured reticular pattern that reaches to the sides of the throat and on the lower lip turns into a banded pattern. On the underside of the stomach and chest, there are pale, reddish brown transverse bands on a light background. The underside of the tail and inside of the legs are a creamish yellow. In young animals, the tail is more-or-less a uniform dark colour. Only when they become adults does the lower one-third become uniform yellow. The tip of the tail in this species is especially sensitive and will die-off easily (Horn & Schürer, 1978).
Habitat: Stony, rocky areas.
Lifestyle, husbandry, and reproduction: These timid and agile monitors inhabit burrows, crevices, or hiding places below boulders. Although they are also diurnal, they are mainly encountered after sunset when hunting for insects. To house this species, a vivarium of at least 150 × 50 × 50 cm is required. A large water container should always be available. This may safely cover half the floor area of the vivarium since these monitors will make frequent use of it for bathing. The other half of the base

Varanus glebopalma

of the vivarium should consist of loosely stacked, flat sandstone rocks. According to Horn & Schürer (1978), this monitor feeds on small mice, lizards, geckoes, and crickets. Nothing is known about their breeding habits.

Varanus griseus (Daudin, 1803)
Desert Monitor

Distribution: From North Africa through southwest Asia to northern India.
Description: Desert Monitors may reach a total length of 130 to 150 cm. Attached to the short pointed head there is an elongated sturdy body. The slitlike nostrils lie at an angle. The extremely long tail, which in cross section is round, is not keeled. The back and tail are greyish yellow to brown with dark brown transverse bands. A dark brown stripe runs from each of the upper and lower eyelids to the sides of the neck.
Habitat: Dry, hot, sandy deserts and steppe regions with sparse vegetation.
Lifestyle, husbandry, and reproduction: In the wild during spring, these monitors are active during daylight hours and will hunt prey in the morning, afternoon, and evening. During the hot summer months, the days are spent in burrows that they dig themselves, or those of other animals. Juveniles usually hide beneath rocks. In the wild these lizards feed on vertebrates, small lizards, snakes, and birds eggs.

Desert Monitors mate in spring. In June and July the female will lay from 5 to 20 eggs in a hole in the ground. The eggs measure around 30 × 20 mm and are immediately covered by the female. The young hatch in autumn. From the end of October until the end of March, these monitors hibernate in their burrows. This species may be kept in the same way as *Varanus exanthematicus*, but during the year it is imperative that they are subjected to temperature variations.

Varanus indicus (Daudin, 1802)
Mangrove Monitor

Distribution: On Celebes, Timor, the Moluccas, the Bismarck, and Solomon Islands, on numerous south seas islands; New Guinea, and in northern Australia.
Description: With its long neck, elongated body, and long, laterally compressed tail, this imposing 150 cm monitor is particularly well adapted to a life in water. In addition, on the upper side of the tail there is an additional keel. *Varanus indicus* has a pointed head with the nostrils on the tip of the snout. Large rectangular scales cover the top of the head. The upper side of the body is brown to black; yellowish to white flecks adorn the flanks. The underside is white or cream.
Habitat: Near water in rain forests, in rivers, and amongst mangroves near the coast.
Lifestyle, husbandry, and reproduction: The Mangrove Monitor always lives near water. Here the lizard will often rest on branches overhanging the water. At the slightest disturbance or hint of danger, it drops into the water and submerges immediately. The Mangrove Monitor will also occasionally venture into the sea. This lizard is not only an excellent swimmer but is also a very adept climber and runner. In the wild, this monitor feeds on crayfish, insects, and fish. Larger specimens will also eat snakes, young crocodiles, birds, and mammals.

There is scant information about breeding in the wild. In one clutch there were seven eggs measuring around 58 × 28 mm. Newly hatched young are around 25 cm in length.

For these monitors, the largest possible vivarium with a very large water container or section is essential. A stout branch should be firmly fixed above the water. The temperature should fluctuate between 25 and 32 °C, and below a spotlight should be locally raised to 40 °C. The temperature of the water should be kept at 25 to 27 °C. Plants are not necessary in the vivarium; they would invariably be destroyed by the lizard. A few hanging plants placed well out of the reach of the lizard may be installed for decorative purposes. The land part of the vivarium should include a box containing a moist mixture of sand and peat. An artificial cave or a hollow tree trunk will serve as an admirable hiding place and refuge.

Varanus karlschmidtii Mertens, 1951
Schmidt's Monitor

Distribution: Papua New Guinea, in the tropical rain forest zone (Sepik River).
Description: This animal reaches a total length of around 120 cm. According to their origins, they may be very differently coloured. In some species the ground colour

Varanus karlschmidtii

on the body and tail is dark brown to black with numerous white to yellowish spots. On the blue-green tail there is an insignificant, dull-yellowish, transverse band. The neck region tends to be a yellowish tone. The cranium is greyish blue. Throat, lower jaw, nose, and eye regions are a flesh-red colour, changing from pale pink to reddish yellow on the neck. The tip of the bright red tongue is white. Other specimens may have a black to greyish black ground colour with yellowish green spotting. Four transverse bands cover the front part of the body. In these animals the neck region is dark brown to brown-black with individual white scales interspaced with intermittent orange-yellow scales. The pale red tongue of these specimens is black on the tip (Horn, 1977).

Habitat: Dense rain forests and swamps.

Lifestyle, husbandry, and reproduction: This monitor is exclusively aquatic. Horn keeps them in a vivarium of 150 × 80 × 100 cm with a substrate of loam and peat which is kept moist by daily spraying. Below the substrate there is a 60-W heating cable, whilst at the top of the vivarium there is a 150-W Vita-Light™ tube. Several stout branches are installed for climbing and a large hollow tree trunk serves as a hiding place. It is imperative that the vivarium for this species is sprayed daily. During the day, the temperature should be 28 to 32 °C, falling at night to 22 to 24 °C. This species enjoys basking beneath a spotlight where the temperatures are considerably higher. The water should be kept at 26 to 28 °C. Horn's Monitors of this species prefer to eat frogs, but will also eat locusts, live mice, and dead chicks, all of which are coated with a vitamin/mineral preparation. *V. karlschmidtii* will also eat carp. Animals such as frogs however should never be taken from the wild for feeding purposes.

Varanus mertensi Glauert, 1951
Mertens' Water Monitor

Distribution: Probably restricted to the area around Lake Torrens in Australia.
Description: This attractive monitor rarely reaches a length of more than 120 cm. The back is dark brown with the extremities appearing even darker. The back and first one-third of the tail have yellow spots that form 10 to 12 poorly defined transverse bands. The temporal area and sides of the snout are a yellowish colour whilst the throat is lemon yellow. Starting from the corner of the mouth, there is a short blue-grey stripe that changes to a band at the throat and becomes blurred as it reaches the neck. The eyelids are yellow and the iris blue-grey whilst the tongue is blue-black. The underside of the body and tail are yellow to light yellow with blurred transverse bands that are partially interrupted by yellow sections. The transverse bands reach from the front limbs to the first one-third of the tail (Schürer & Horn, 1976).
Habitat: In the direct vicinity of water.
Lifestyle, husbandry, and reproduction: *Varanus mertensi* lives in the vicinity of water. This monitor is rarely seen in captivity. A male of this species was once kept in a vivarium of 200 × 80 × 100 cm. The 20-cm-deep water container was made from aluminium. Two-thirds of the ground area was covered with a mixture of peat and loam and one-third with rough gravel. A hollow oak stump and a stout 3 m long forked branch completed the furnishings. The water temperature varied between 26 and 30 °C, whilst the air temperature was 24 to 26 °C. Beneath a spotlight the temperature ranged from 26 to 38 °C. Mertens' Water Monitor prefers to eat fish ranging from 10 to 16 cm, but will also eat juvenile mice and dead chicks. In the case of Brotzler (quoted in Schürer & Horn, 1976), a female laid 13 eggs, 7 of which were deformed. The first of the five young hatched after 182 days and the last after 217 days. After hatching, the 25.2 to 27 cm long young weighed between 23.6 and 27.9 g. After 5 or 6 days they ate small fish, crickets, and young mice coated in egg yolk.

Eidenmüller (1990) successfully kept a pair in an aquarium measuring 150 × 60

× 40 cm, with a 60 cm high glass superstructure. Sliding doors in the front panel of this superstructure allowed access. The water in the base of the aquarium was 35 cm deep. The land section was polystyrene sheets covered with sand. The adhesive used was epoxy resin. This gave the impression that the land section was sandstone and also kept the monitors' claws trim. A strong willow branch protruded from the water, which was kept at 28 °C. An internal filter ensured that the water was kept clean. Two mercury vapour lamps, each of 50-W, kept the vivarium brightly illuminated, whilst their starting motors served as ground heating. Additional heating was provided by a Concentra™ spotlight. The lizards were able to reach a second, dark section of the vivarium through a flexible tube 10 cm in diameter. The second section contained a 30 cm deep layer of sand and was kept at a temperature of 27 to 30 °C. This chamber was used for sleeping and egg laying. On September 3, 1988, Eidenmüller saw several copulations, most of which took place in the water. Three weeks later, the female laid six soft-shelled eggs in the sand in the second chamber. The eggs, which measured 6 to 6.5 × ca. 3 cm, were placed in incubation boxes containing vermiculite or Perlite™. The first of the young hatched after 265 days and the sixth after 316 days. The incubation temperature fluctuated between 27 and 28 °C whilst the atmospheric humidity was around 95%. The young had total lengths from 29.4 to 31 cm and weighed between 25.9 and 29.5 g. Apart from being more intensely coloured, the young were identical to their parents in both colouring and markings. After 5 days they took their first meal of pink mice, crickets, grasshoppers, and live-bearing tooth carp (Platties). Before being given to the monitors, the mice and crickets were liberally dusted with a vitamin/mineral preparation. The adults were fed for the most part on large insects, mice, and fish.

Varanus niloticus (Linnaeus, 1758)
Nile Monitor

Distribution: Africa, with the exception of the dry areas of the Sahara Desert and the Mediterranean coastal strip. The Nile Monitor is absent from southwest Africa.
Description: The Nile Monitor may reach a maximum length of over 2 m. Its head appears to be short, flat, and tapered towards the snout. On the upper side of the long tail there is a keel. The scales on top of the head and on the limbs are small, narrow, and overlapped in the manner of roof tiles. At the rear of the body the scales are larger, elongated, rectangular, and unkeeled. The dark olive to blackish green upper side has scattered small white spots and large white flecks forming transverse bands. As the animal becomes older, the markings gradually fade. The upper sides of the limbs have yellow flecks, whilst the tail has black and yellow bands. Yellow lines run from the head to the neck, where they form a horseshoe-shaped marking. The underside is yellow with more-or-less black bands.
Habitat: In the vicinity of standing and flowing water.

Varanus niloticus

Lifestyle, husbandry, and reproduction: The Nile Monitor always lives in the vicinity of water. They may be frequently found basking on branches or rocky outcrops hanging above water. At the first sign of danger, they will drop into the water and swim away quickly with the legs tucked neatly alongside the body. Whilst swimming, the propulsion and directional control are taken over by the tail. This Monitor is also an excellent diver. When threatened, the Nile Monitor inflates its throat and hisses loudly. When molested it will lash out with the tail and inflict a very severe bite without any warning. As food they will eat all manner of aquatic life, including mussels, snails, crayfish, other lizards, turtles, birds and their eggs, small mammals, and even crocodile eggs. When young they feed mainly on insects.

Females lay 40 to 60 eggs measuring around 54 to 60 mm in length and 30 to 40 mm in width. Eggs are usually laid in termite mounds. The hole that they use to do this is quickly sealed again by the termites. Protected in this way, the young develop within 9 to 10 months. After hatching, the young break out from the termite mound and search for water.

This species should be kept in the largest possible vivarium with the same furnishings as described for other monitors.

Varanus prasinus (Schlegel, 1839)
Emerald Monitor

Distribution: New Guinea and neighbouring islands.
Description: The Emerald Monitor reaches a total length of around 70 cm and has a

Varanus prasinus

short head with a long pointed snout. The nostrils lie on the sides of the head behind the snout. The upper side of the head and neck have smooth scales, whilst those on the body are keeled. The light green upper side of the body has small black spots creating a reticulated pattern or narrow transverse bands. The underside may be a whitish, yellowish, or greenish colour.
Habitat: Rain forests and mangrove thickets.
Lifestyle, husbandry, and reproduction: Very little is known about the lifestyle of this species in the wild. This monitor prefers to climb trees using the tail as a prehensile organ.

In captivity they will eat chicks, mice, and occasionally fruit. Emerald Monitors should be kept in spacious vivaria with large water containers and several climbing branches, some of which should project above the water. A hollow tree stump will serve ideally as refuge. The temperature should be between 28 and 30 °C, even higher beneath a spotlight where the lizards will spend many hours basking, especially in the early hours of the morning. The water temperature should be kept at 25 to 27 °C.

Varanus rudicollis (Gray, 1845)
Rough-Necked Monitor

Distribution: From southern Burma through the Malayan Peninsula, the Rhio Archipelago to Sumatra, Bangka, and Borneo.

Description: The common name is derived from the greatly enlarged scales on the neck. This species reaches a total length of 100 to 120 cm, in rare cases even up to 160 cm. The upper side of the head is almost black, whilst the neck is yellow with three dark stripes. At the front of the yellowish back there are two wide black transverse bands. The legs have yellow flecks, whilst the tail has broad black and yellow transverse bands. Adult specimens are almost uniform black.
Habitat: In dense jungles on trees. Also found in the mangroves along small coastal rivers.
Lifestyle, husbandry, and reproduction: Rough-Necked Monitors are expert tree climbers. They are extremely timid creatures, a phenomenon that does not change even after a long period in captivity. The young usually hide amongst fallen leaves and feed mainly on insects. In the vivarium, this monitor will eat chicks, mice, small rats, and crickets. Juveniles will take small fish and later nestling mice and locusts (Horn & Peters, 1982).

A female measuring 115.5 cm that had been newly imported by a dealer laid 13 eggs on January 17, 1981. These weighed on average 60.4 g and measured from 57 mm in length and 44 mm in width. Five of these eggs proved to be infertile. During incubation the temperature fluctuated between 28 and 30 °C. Occasionally, after spraying, the humidity rose to 100%. The first of the young hatched on July 15, 1981, with the remainder following soon afterwards. The hatchlings measured on average 25.2 cm and had an average weight of 20.6 g (Horn & Peters, 1982). Another female laid four fully formed eggs measuring 55 to 60 mm in length and 31 to 32 mm in width. These eggs, however, proved to be infertile.

Varanus salvator (Laurenti, 1768)
Water Monitor

Distribution: Southeast Asia; in the north to southern China, Hainan Island in the south and east; in Sri Lanka, the Andamans, Nicobars, Philippines, in the Indo-Australian Archipelago to Sulawesi (Celebes) and Wetar. There are several geographic subspecies.
Description: The Water Monitor may reach a SVL of 104 cm and a tail length of 132 cm. Rotter (1963) even mentions animals with a total length of 300 cm. On the brown to black upper side, this monitor has yellow spots or flecks in transverse rows. The snout is lighter; there are dark transverse bands on the lips and chin. A black stripe, edged in yellow stripes, starts from the eye and runs down to the neck. In juveniles, the black and yellow patterning is especially pronounced. The scales on the head are slightly enlarged whilst those on the neck are small, oval, and keeled. The scales on the underside are also keeled. The tail is strongly laterally compressed with keeled scales and a flat double comb. The limbs have only relatively short claws.
Habitat: In the vicinity of water, usually on banks.

Varanus salvator (left), Varanus spenceri (right)

Lifestyle, husbandry, and reproduction: On the island of Negros in the Philippines, the natives hunt this animal relentlessly for the delicious flesh and for the skin. The animal is so extensively hunted that its numbers have been greatly reduced. They are also slaughtered to protect domestic poultry. Some natives of Negros live exclusively from hunting Water Monitors and the sale of their skins which are exported to Japan and western countries (Gaulke, 1986b).

The Water Monitor is an excellent swimmer and diver. Because of its enormous size, this monitor should only be kept by zoological gardens and reptile houses. The Water Monitor requires an enormous vivarium with a very large section of water. Temperatures should be from 22 to 28 °C and even higher beneath a spotlight. An artificial cave will frequently be used by these animals, whilst stout branches and tree trunks will be appreciated for climbing. The floor and water section should be easy to clean. A large box containing a mixture of sand and peat should be placed in the vivarium for egg laying. Water Monitors will eat fish, pieces of meat, snails, mice, and dead chicks.

Katzler (1973) received a newly-laid clutch of Water Monitor eggs containing 14 fairly large hard-shelled eggs. On average, the eggs measured 64 to 71 mm in length, 34 to 37 mm in width and weighed 46 to 50 g. They were placed in a plastic box containing semi-moist peat and then covered with fine mesh and a sheet of glass. Incubated at temperatures between 28 and 32 °C, the first hatchling appeared after 241 days. The last of the 11 young only appeared after an additional 86 days.

Lizards

Varanus spenceri Lucas & Frost, 1903
Spencer's Monitor

Distribution: Central parts of northern Australia and Queensland.
Description: This lizard reaches a length of around 140 cm, whereby the tail is equal to the SVL. The skin is extremely rough and the greyish brown ground colour is somewhat reminiscent of *Varanus griseus*. On the neck, back, and tail there are 24 light transverse bands.
Habitat: Dry, sandy bushland.
Lifestyle, husbandry, and reproduction: The temperament of this lizard is very different from that of *Varanus griseus* and they quickly become accustomed to life in captivity. This is a very peaceful lizard which only rarely digs. They will readily eat mice and small rats (Peters, 1968).

From a pair that were caught in September 1968 whilst mating in the area of Tennants Creek in northern Australia, a female laid a clutch of 18 eggs on November 11, 1968. Each of the eggs measured 52×36 mm. Because there was no way to bury the eggs, they were scattered around the floor of the vivarium which was covered with peat (Peters, 1969, 1985). Four of the eggs were obviously infertile, whilst the remaining 14 were incubated under different conditions:

1. After having been exposed to ultraviolet light for 4 minutes to prevent the formation of mould, two eggs were placed on moist toilet tissue in a sterilised glass container.
2. Four eggs were placed in a sand-filled earthenware container that was then placed in a second sand-filled plastic container. Only the sand surrounding the earthenware container was kept moist during incubation.
3. Three eggs were buried 2 to 3 cm deep in a sand-filled plastic container.
4. Three eggs as in (3) but in a special container.
5. Two eggs as in (1).

Each of the containers containing eggs was placed in a darkened aquarium at a temperature of 29 °C. One moulded egg was removed after 4 days and the remaining eggs, except those in container (1) were dusted with an antimycoticum. This powder contained 2% pentyoxybenzamid and 0.5% hexachlorophene. After being treated in this way, none of the other eggs developed mould. After 80 days, the eggs in container (5) had two-thirds dried-out; they were opened and two live embryos of 78 and 83 mm were found. Incubation lasted a total of around 4 months until the 11 young monitors had hatched within 7 days of one another. The first hatchling measured 222 mm, consisting of head 22 mm, body 102 mm, and tail 98 mm. The young were considerably different from the parents in that they had very striking colouration. The velvet brown back had eight or nine broad yellow bands, and on the upper side of the neck there were two or three V-shaped yellow stripes. The underside was white. The hatchlings fed after 4 days, taking young mice and mealworms. Three of the hatchlings doubled their size in only 6 weeks. Additional food in the form of raw egg and strips of beef heart was given to these three hatchlings.

Varanus storri Mertens, 1966
Storr's Monitor, Storr's Dwarf Monitor

Distribution: There are two subspecies: The nominate form lives in Queensland, *V. storri ocreatus* in the area of Lake Argyl in Western Australia and in the adjoining Northern Territories.
Description: This monitor reaches a total length of up to 44 cm. Its upper side is pale to dark reddish brown with scattered dark brown to black scales. These sometimes form a weak reticulated pattern. The underside of the head and sides of the neck are white or cream. Storr's Monitor has small, irregular and smooth scalation on the head. The nostrils are rectangular. Because of this, this species is somewhat reminiscent of *V. primordius*. As opposed to *V. primordius* however, *V. storri* has 70 scales at the centre of the body. The sexes are not easily distinguished. In both sexes there are small spikes near the root of the tail. If the animal is placed on a smooth surface and viewed from above, the spikes of the male are easy to see; those of the female less so.
Habitat: Light dry forests, in small burrows that they dig themselves beneath piles of rocks. The ground in these places is usually clayey with sparse vegetation. This species has no need for trees.
Lifestyle, husbandry, and reproduction: In the natural habitat of *Varanus storri*, Peters found that 80% of the trees had died. Spear grass and rocks were the main features of the landscape. In an area of 1 3/4 km² he saw only 13 adult males and 7 females, 5 of which were gravid. The preponderance of males and lack of juveniles from the previous year was particularly noteworthy. The specimens observed measured between 170 and 295 mm. They always remained solitary in tunnels which they had dug themselves beneath rocks. The tunnels were U-shaped with a larger chamber at the end. Further tunnels and the tracks of other monitors indicated a population of 45 to 50 specimens.

These lizards were easily found before daybreak because from 0900 to 1300 h the tunnels were all empty as the animals were searching for food. During the heat of midday the animals retreated to their tunnels where they remained until late afternoon.

When attempting to establish a breeding pair, this is best left to the animals themselves, because both males and females will fight amongst themselves, whilst another combination of partners will be quite harmonious (Flugi, 1990).

This species requires a vivarium of 70 × 80 × 75 cm or larger. The rear wall should be covered with pieces of rock. A suitable substrate is a 10 cm layer of clayey earth below which a heating cable has been installed. Piles of rock and a gnarled branch should be placed on the floor of the vivarium. Storr's Monitor requires a humidity of 60 to 65% and ground heat of 30 to 40 °C (Stirnberg & Horn, 1981).

In the wild, Storr's Monitor feeds on insects, spiders, small geckoes, and juvenile agamas inhabiting the same area. In the vivarium they will eat house crickets, field crickets, juvenile locusts, and nestling mice.

The clutch usually contains two to four eggs. Even at a total length of only 29.5 cm, a female produced eggs. After this female had produced a number of infertile eggs, a hatchling measuring 12.3 cm (SVL 5.4 cm, tail 6.9 cm) was unexpectedly found in the vivarium. It was reared on small crickets and small waxmoth larvae. After only 3 months this juvenile had already reached a total length of 22.3 cm (SVL 9.1 cm, tail 13.2 cm).

The spindlelike eggs of Storr's Monitor are around 2.5 cm long and 1 cm wide. With high humidity and temperatures around 27 to 29 °C, they take between 103 and 110 days to hatch. Vermiculite has proved to be the most suitable incubation medium. Immediately after hatching the young measure SVL 4.8 to 5.6 cm and tail 6.8 to 8.2 cm (Eidenmüller & Horn, 1985). Suitable foods for rearing this species are small field crickets, waxmoths and their larvae, and newly caught grasshoppers. After a while it is best to house the young individually.

Varanus timorensis (Gray, 1831)
Timor Monitor

Distribution: Timor and neighbouring islands.
Description: This species reaches a total length of 60 cm. Its grey to black upper side has white or yellowish flecks and spots. The patterning of these animals is quite variable. In some specimens the eye spots give a leopardlike reticulated appearance, whilst in others they are arranged in rows. These spots can sometimes however be only barely visible. Usually a pale stripe delineates the white underside. The limbs are speckled with white, whilst the tail is greyish black with either regular or irregular rings of white scales. The scales on the head are small and smooth.
Habitat: On trees away from forests.
Lifestyle, husbandry, and reproduction: A vivarium measuring 150 × 50 × 50 cm is suitable for a pair of these Monitors. Sand or cat litter is a suitable substrate. A large drinking/bathing bowl is one of the most important features of the vivarium for this species. During the day, the temperature should be between 28 and 30 °C, falling at night to 20 to 25 °C. During the day, part of the floor of the vivarium should be heated to 35 °C.

Once or twice weekly, the animals are given house crickets, field crickets, young mice, nestling mice, fish, or eggs. This monitor has been bred several times in captivity. Usually, the female buries between 2 and 12 eggs in the substrate. These

Varanus timorensis

should be bedded in vermiculite and incubated at temperatures between 27 and 39 °C with high humidity. Under these conditions, the young hatch after 119 to 127 days measuring 13 to 15 cm. They are easy to rear when fed on a variety of small insects (Eidenmüller, 1986; Behrmann, 1981; Rese, 1983; Sauterau & de Bitter, 1980).

Varanus similis Mertens, 1958
Spotted Tree Monitor

Distribution: Australia, from northeastern Western Australia through the Northern Territories to Queensland and to the north of New South Wales. Also on the islands in the Torres Strait and in New Guinea.
Description: *Varanus similis* reaches a total length of up to 60 cm. In males, the postanal scales are elongated to form spines and are arranged in groups (a very uncertain characteristic!). The males also have a larger head and thicker tail root than females. The round tail is around 1.5 times as long as the SVL and the legs are relatively short. The colour of these monitors is quite variable. The upper side may be yellowish grey to deep black. In lighter specimens, darker scales may form eyespots. In dark specimens, this pattern is reversed. On the body, the eyespots often form transverse bands. They start at the neck and tail but elsewhere appear as individual

Lizards

spots. Occasionally there are individual brown or reddish brown spots between the rows of eyespots along the back. *Varanus similis* is considered by many herpetologists to be a subspecies of *Varanus timorensis*.

Habitat: A tree dweller.

Lifestyle, husbandry, and reproduction: This species lives mainly on trees and may often be found beneath loose bark. In an area of 50 m² Schmida (1971) found 14 specimens in the 30 hollow tree trunks that he examined. This species has furthermore evolved to be a follower of civilisation. Care and husbandry are as described for *Varanus timorensis*. These animals are also found on the ground when they are searching for food.

In 1972, Rüegg (1974) obtained a pair of *V. similis*, and in November and December 1972 twice observed copulation. During the night of January 19/20, the female laid five soft-shelled eggs beneath the roots of a plant. Rüegg incubated three of these in an earthenware container holding a mixture of sand and peat. The incubation temperature varied from 28 to 31 °C, and the relative atmospheric humidity was 80%. On March 17, 1973, one egg had developed mould and had to be removed. Two eggs that had been left where they had been laid were the victims of an accident. On the morning of June 8, the head of the first of the two young was seen protruding from the egg. It only completely emerged the following day. Sadly it lived only a very short time. After an incubation period of 140 days, the second egg hatched. The hatching had a yellow throat, whilst over the entire body there were transverse bands of light eyespots on a dark background. From the third day after hatching, the young *V. similis* took crickets and later waxmoths and their larvae, Guppies, and strips of beef heart muscle. By September 12, its length had already increased to 198 mm.

Eidenmüller & Wicker (1991) also report the successful breeding of this species. Their monitors were kept in a vivarium measuring 105 × 70 × 60 cm (L×W×H). The rear and side walls were clad in polystyrene sheets affixed with epoxy resin. The substrate was sand. Some stout branches and a hiding box measuring 40 × 20 × 20 cm completed the furnishings. The hiding box contained moist sand. Beneath a spotlight a localised temperature of 42 °C prevailed. In summer the lighting was switched on for 14 hours daily and in winter for 9 hours. The same cycle applied to the spotlight. At night the temperature fell to around 20 °C. In this way the keepers imitated the annual cycle of the southern hemisphere in an attempt to encourage the natural sexual behaviour of the animals.

Copulation took place in November, 1988 and eggs were laid during the following January, as had been the case the previous year. The first three of the eggs were incubated in moist vermiculite at 26.5 to 28.5 °C. After 134 days the first egg hatched. The hatchling had a SVL of 6.5 cm and a tail 11 cm long.

Of the six eggs laid in 1989, three were infertile. They were incubated at 27 to 28 °C. The three young hatched after 132 to 136 days. They each had a total length of 14.8 cm and grew much more slowly than the specimens that had hatched previously.

Varanus tristis Schlegel, 1839
Black-Headed Monitor

Distribution: Northern Territories, Queensland, Western Australia. The nominate form lives in west to central Australia, *Varanus tristis orientalis* in a region stretching from Queensland to northern Australia.

Description: This species is divided into two subspecies and nowadays one speaks only of a "flecked" and a "black-headed" form (Wilson & Knowles, 1988; quoted in Eidenmüller, 1991), because the ranges of the two overlap producing numerous intergrades. *V. tristis* may reach a total length of 80 cm. On the pale blue, dark brown, or black upper side, there are many dirty-white or white-bordered eyespots with a dark centre. In adults, these spots are very weak or even absent. The limbs and tail have cream-coloured spots on a black background. On the tail these may form narrow rings. The scales on the head are small and smooth. In cross section the tail is almost round.

Habitat: Approximately the same as for *Varanus timorensis*; in trees, rock crevices, or piles of rocks in the drier parts of its distribution range.

Lifestyle, husbandry, and reproduction: A vivarium of $150 \times 70 \times 100$ cm is suitable for a pair of this species. The remaining furnishings should consist of a tree stump or tree root and pieces of circular cork bark. An artificial cave will also be greatly appreciated. In all other respects, this species should be kept in the same way as *Varanus timorensis*.

Females lay eggs around 6 weeks after mating, usually in moist sand in the hiding box or artificial cave. If the eggs are incubated at 27 to 28 °C and 100% humidity, the young will hatch after 110 to 135 days. The hatchlings take their first food after 2 to 4 days.

In Eidenmüller's case, from the second clutch in one year, the first young hatched after 116 days, whilst the last did not emerge until after 133 days. Two of the hatchlings had spirally coiled tails, whilst the other two had no tail at all! Because the spiral tails proved only to be an inconvenience, they were amputated. The hatchlings fed after 3 to 5 days, taking field crickets. Eidenmüller was fortunate in being able later to breed perfectly formed young which he has now bred into the second generation.

Varanus varius (Shaw, 1790)
Lace Monitor

Distribution: Australia, from Princess Charlotte Bay along the coast of Queensland and New South Wales to Port Philipp Bay in Victoria. In southern Australia there is also a further isolated population.

Description: With a total length of up to 228 cm, the Lace Monitor is the second largest of all Australian lizards. These animals are very differently coloured. On a light brown to blue-black background there are irregular dirty-white, yellow, light yellow, or yellowish green bands. Occasionally the lips, throat, and temporal area have vivid blue spots. The limbs frequently have a yellow pattern of various spots or bands. Often the tail will only be partly banded. Some specimens are even beautifully two-toned (the "*bellii*" phase). In their case, wide yellow-green and black bands cover the entire body.

Habitat: Areas with many trees, above all in rain forests and regions with average to heavy rainfall at altitudes from 800 m above sea level. In some areas the summer temperature may reach 40 °C, whilst winter produces mild frost.

Lifestyle, husbandry, and reproduction: This monitor will often hibernate for several months during the coldest part of the year. However, even on a sunny day in winter they may be found basking. In the tropics, this lizard is active throughout the year. Normally this monitor seeks food on the ground, but at the first sign of danger will seek refuge in a tree. On warm, sunny days, large specimens will maintain only a short safety distance.

In the wild, the Lace Monitor mates in November and December. The eggs are then laid in late December or early January. Their food consists mainly of animals and their products, above all, eggs, but also carrion. In the vivarium *V. varius* should be kept as *V. flavirufus* or *V. tristis* but should be given a much larger water container.

ALLIGATOR LIZARDS
Family Anguidae

The representatives of this family live predominantly in the temperate and subtropical regions of the northern hemisphere. Some species occur in South America, but only a few have reached Asia. The members of this family can be divided into three distinctly identifiable subfamilies. Some herpetologists still consider them to be separate families. The representatives of the subfamily Diploglossinae are only rarely available to herpetologists, and very little is known about their life in the wild.

The body of an Alligator Lizard is greatly elongated and the long tail breaks off easily, only partially regenerating. In this family of lizards can be found not only relicts of originally four-legged species, but also the transitory species to completely legless lizards. Along the sides of the body there are many folds of skin. There may be skin ossification below the scales.

Alligator Lizards and Glass Lizards
Subfamily Gerrhonotinae

Gerrhonotus Wiegmann, 1828
Alligator Lizards

The genus *Gerrhonotus* contains around 15 species with an enormous distribution range which includes the west and north of Central America and North America. The northernmost limit is southern Canada. Almost all species prefer open areas that are not too dry. The body is elongated, the head flat and cone-shaped. The four extremely thin limbs each has five toes. The tail is almost as long as the SVL. In some tree-dwelling species, the tail serves as a prehensile gripping organ. The body is covered with large platelike scales that overlap in the manner of roof tiles. These scales are only absent from the sides of the body.

Gerrhonotus coeruleus Wiegmann, 1828
Northern Alligator Lizard

Distribution: Northwest United States to southwest Canada.
Description: According to subspecies, this lizard may reach a total length of up to 35 cm, whereby the tail is almost double the SVL. Alligator Lizards are usually olive brown with numerous dark spots on the body. These may sometimes form transverse bands. There are dark longitudinal stripes on the underside.
Habitat: Beneath piles of wood or rocks. Moist, cool forest regions up to an altitude of 3200 m above sea level.
Lifestyle, husbandry, and reproduction: This lizard feeds on a variety of arthropods, earthworms, snails, and, occasionally, birds' eggs.

Alligator Lizards require spacious vivaria with several hiding places and climbing branches or tree roots. A suitable substrate is a mixture of forest loam and sand. Two-thirds of the vivarium should be planted and kept slightly moist. A small spotlight above the unplanted part should produce a localised temperature of 35 °C, otherwise temperatures of 22 to 25 °C are quite adequate. The vivarium should be sprayed daily. During the summer months, it is advantageous to house this species in an outdoor vivarium which should be covered if necessary.

Alligator Lizards are usually active during the morning and evening. In the wild, mating takes place in April, at higher altitudes in May or June. Around 7 to 10 weeks after mating, the female gives birth to between 2 and 15 young. On February 10, 1983, Brennecke observed the birth of young. The female lifted a hindleg and an egg-shaped mass was released from the cloaca. The female lowered the hindleg and dragged the young along the ground until it was completely free of the cloaca and the transparent amniotic sac was broken. After around 30 minutes the young lizard began moving, and after a further 30 minutes was running around quite happily.

Lizards

Gerrhonotus coeruleus

After yet another 30 minutes the yolk sac had broken off. Further young followed at half-hourly intervals. All eight young were fawn to dark brown with central stripe. Each of the young measured 25 mm and began eating larvae and small crickets at 3 days old. Two of the young had vertebral deformities and died quickly. The remainder grew at different rates.

Gerrhonotus kingi Gray, 1838
Madrean Alligator Lizard

Distribution: Southeast Arizona and southeast New Mexico, southwards along the Sierra Madre Occidental to Mexico.
Description: This lizard may reach a total length of 31.7 cm, but is usually smaller. They have light orange to pink eyes. There is a distinct fold along the sides of the body. They are brown with crossbands along the back. On the upper lip there are white and black spots, whilst on the underside there are spots and flecks. Juveniles have light crossbanding.
Habitat: Dry grassland, open woodland at altitudes of 1,500 to 2,100 m above sea level.
Lifestyle, husbandry, and reproduction: As for *Gerrhonotus coeruleus*. This is a

Gerrhonotus kingi

ground-dwelling species. In the wild, females lay between 9 and 12 eggs during June or July.

Gerrhonotus multicarinatus (Blainville, 1835)
Southern Alligator Lizard

Distribution: Southwest United States and northern Mexico to Baja California.
Description: There are several subspecies, the largest of which reaches a length of 40 cm. They are usually yellowish grey to brown and have dark, white-edged transverse bands along the body. These become narrower towards the tail. In the centre of the scales on the belly there are dark stripes. Juveniles have a light band on the back.
Habitat: Open woodland, moist areas, mainly in oak woodland, and at the foot of hilly regions.
Lifestyle, husbandry, and reproduction: The Southern Alligator Lizard is also predominantly active during the day. When searching for food they will occasionally climb amongst bushes. When climbing they use their tail as a prehensile organ to cling to branches. If threatened they emit a foul smelling liquid from the cloaca and inflict a very painful bite. During the warmer months of the year, females lay two to three clutches, each containing 1 to 41 eggs. However, a more usual number is 12.

Langerwerf (1979) kept his Alligator Lizards in an outdoor enclosure measuring 10 m^2. Males are very aggressive towards one another. It is therefore best to keep

Lizards

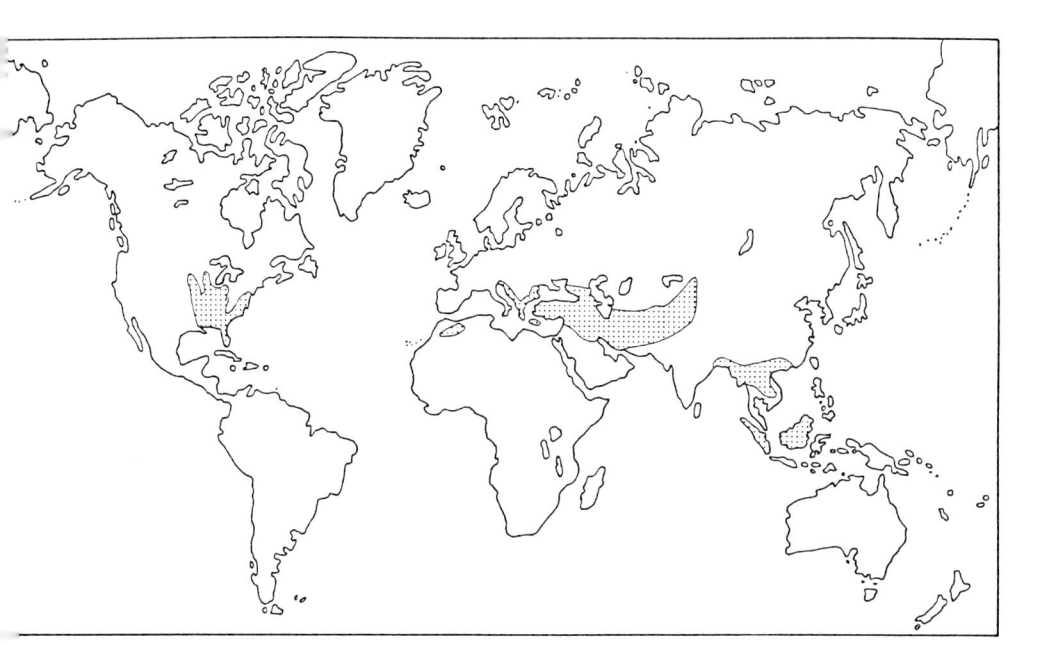

Distribution ranges of *Ophisaurus*

these animals in pairs, although one male may be kept with two or three females. The 17 × 10 mm eggs take around 40 days to hatch if incubated at 30 °C.

Ophisaurus Daudin, 1803
Glass Lizards

Glass Lizards are at home in southeast Europe, North Africa, western and southern India and in southern North America. They live in very divergent habitats such as bushy steppes and moist forests. The largest species may reach a length of over 1 m. Limbs are almost absent or are limited to the relicts of hindlegs. Unlike snakes which are very supple, the body of these lizards is relatively stiff as a result of the large scales. The cone-shaped head is rather blunt. The typical folds along the sides of the body are quite pronounced in this subfamily.

Ophisaurus apodus (Pallas, 1775)
Scheltopusik, European Glass Lizard

Distribution: The main distribution range stretches from the northern Caucasus through Asia Minor and the Middle East into the Central Asian part of the former Soviet Union. In Europe, the Scheltopusik lives on the Balkan Peninsula as well as in isolated populations in Crimea and on the Caucasian Black Sea coast.

Ophisaurus apodus (left), Ophisaurus ventralis (right)

Description: These powerful legless lizards may reach a total length of 125 cm, of which three-quarters is the tail. When young, the body scales are keeled but with increasing age they gradually become smooth. In adults, only the scales on the tail remain keeled. Minute relicts of the rear limbs are present. With the exception of the lighter head, *Ophisaurus* are relatively uniformly coloured. The underside is somewhat lighter. Reddish flecks may be present. Specimens from the Caucasus and Transcaucasia usually have a mosaiclike pattern caused by irregular patches of lighter scales on the back and flanks. Juveniles have irregular transverse bands which they lose at a length of 15 to 17.7 cm at the age of 2 to 3 years.

Habitat: This species prefers sunny places with medium vegetation and sunny slopes.

Lifestyle, husbandry, and reproduction: Scheltopusiks are ground dwellers, usually active during the day. They should be kept in the largest possible vivarium with a water bowl, climbing branches, and a daytime temperature of 26 °C. A localised temperature of 30 °C is needed beneath a spotlight. At night the temperature may safely drop by 5 to 10 °C. In summer, this lizard may be kept in an outdoor enclosure.

The Scheltopusik feeds mainly on snails and slugs, arthropods, small mammals, young lizards, and birds eggs. In captivity it may also be given strips of beef heart. The vivarium should be sprayed daily. According to their origins, hibernation takes place from October or November until March or April. The female lays 6 to 12 soft-shelled, elongatedly oval white eggs after a gravidity of some 10 weeks. The eggs are laid beneath rocks or buried in the ground. The female will frequently guard her eggs by coiling around them. The young, which measure more than 10 cm, hatch after 45 to 55 days. When sloughing, the old skin is rolled together as a ring and rubbed off over the tail.

In Bohn's case (1987), a female buried herself in the substrate from the middle until the end of July. When Bohn removed the animal on July 30, he found a clutch of seven eggs. They lay in a dry place and measured 4.0 to 4.5 × 2.3 to 2.5 cm. The eggs were transfered to an incubator and kept at temperatures of 28 to 29 °C during the day, falling at night to 24 to 25 °C. On September 15, the heads of three young Scheltopusiks protruded from slits in the eggs, which they left during the following day. One of the signs of imminent hatching was that the eggs appeared to have collapsed and had flecks. The remaining eggs proved to be infertile or contained dead embryos. The eggs were in the incubator for a total of 48 days, but it is not known when they were laid. From the fourth day after hatching, the young ate minced beef heart and small slugs from forceps. The adults were placed in a hibernation box on November 3 and the young in a similar box on December 2, 1985. Each box contained a mixture of forest loam, leaf litter and peat and had a fine mesh cover. With temperatures of 7 to 8 °C the hibernation of the adults ended on March 1, 1986. Because one of the young had died, the remainder were removed from hibernation on January 20. After 14 months the young had reached a length of 45 cm.

Ophisaurus ventralis (Linnaeus, 1766)
Eastern Glass Lizard

Distribution: United States coastal regions from North Carolina through Florida to eastern Louisiana.
Description: The Eastern Glass Lizard reaches a total length of 90 to 108.3 cm. Its upper side is olive brown. On the tail there are individual light green spots. At the edges of the dorsal scales there are white flecks.
Habitat: Moist woodlands and fields.
Lifestyle, husbandry, and reproduction: The diurnally active Glass Lizard prefers shaded areas. During the early hours of the morning, these animals may often be found at the sides of roads running through moist areas. The tail breaks off very easily and specimens with regenerated tails are frequently found.

During the period between spring and summer, the females lay between 8 and 17 eggs from which the young hatch after around 2 months. During this time the female remains near the eggs to guard them. The husbandry requirements are identical to those described for *Ophisaurus apodus*.

Slow Worm
Subfamily Anguinae

The Slow Worm is the only representative of the subfamily Anguinae and the genus *Anguis*. It is differentiated from other members of the family by its pointed teeth

Anguis fragilis

and a trace of a hinged joint on the lower jaw. In eastern specimens, an ear opening is still present. There are no external traces of limbs and the tail is exceptionally prone to breaking off. When it does break off, the tail will only regenerate as a stump.

Anguis fragilis Linnaeus, 1758
Slow Worm

Distribution: Throughout Europe with the exception of Ireland, northern Scandinavia, and the greater part of the Iberian Peninsula. In the east, the Slow Worm may be found as far as the Urals and the Caspian Sea, in Asia Minor, parts of the Middle East and northwest Africa. The Slow Worm may be found up to an altitude of 2,000 m above sea level.

Description: The Slow Worm may reach a length of 40 to 45 cm, of which two-thirds is tail. In cross section, the smooth body is round and covered with regularly arranged smooth scales. The basic colour of the upper side is very variable. It may be light grey, grey-brown, brown, bronze, or copper-coloured. The underside is usually a blackish grey to blue-grey. The sides of the body are usually lighter than the back, which may have longitudinal stripes or rows of spots and flecks. Animals with dark blue spots are frequently found in the eastern part of the distribution range. Females differ from males by having a dark stripe along the centre of the back and more sharply defined darker colouring on the flanks. Juveniles are much more brightly coloured and on the upper side are often a gold or silver colour with vivid black stripes along the back. The sides of the body and underside are a very deep black.

Habitat: Fields, heathlands, fallow land, gardens, deciduous forests.

Lifestyle, husbandry, and reproduction: In the wild, the Slow Worm leads a very secretive lifestyle. They are usually found by turning over rocks, leaf litter, or pieces of wood. After rainfall, they may be found above ground searching for slugs, earthworms, or insect larvae.

Slow Worms usually hibernate in groups in frost-free holes in the ground, beneath tree roots, or in old abandoned tunnels of other animals. In central Europe, hibernation lasts from October until April or even early May. Mating takes place immediately after hibernation. Females are gravid for around 3 months, and between June and August give birth to between 8 and 12 live young measuring around 7 to 9 cm. Immediately after being born, the young struggle to free themselves from the amniotic sac. In most countries of Europe, the keeping of Slow Worms in captivity is legally prohibited for conservation reasons, and only animals that are ill or injured may be kept with the relevant permit or licence.

Bregulla (1984) kept several specimens in a vivarium with a ground area of 80 × 40 cm. The substrate was similar to that found in their natural habitat and consisted of a 20 cm layer of forest loam which was kept constantly moist. Soon after being installed in the vivarium, the Slow Worms had constructed a series of tunnels and were only very rarely seen. Feeding the creatures was also difficult and it could only be hoped that they found the earthworms and slugs that were introduced into the vivarium. In some cases, Slow Worms become so tame in captivity that when the vivarium is opened they will immediately come to the surface to accept food from forceps.

The vivarium must be sprayed daily so that the animals may lick up the resultant water droplets to cover their liquid requirement. A semi-shaded window is an excellent place to stand the vivarium so that it does not become too hot inside. An independent heat source is not necessary. Even after having been kept for long periods in captivity, Slow Worms will still lead a very secretive life. Hibernation is essential.

PLATED LIZARDS
Family Gerrhosauridae

Amongst Plated Lizards one usually finds a powerful, flattened body and large head which is barely set-off from the body. Plated Lizards are somewhat reminiscent of skinks because their large scales are arranged in regular, longitudinal and transverse rows. Sometimes the scales are keeled, but never spiny. A deep fold of skin, covered with fine granular scales, is typical of the Plated Lizards. This allows the body to expand considerably. In *Tracheloptychus*, the fold is limited to the neck region.

Angolosaurus Fitzsimons, 1953
Plated Sand Lizards

The animals of this monotypic genus have a short pointed snout, a cylindrical tail, and a cylindrical body. They have long limbs and a fringe on the side of each toe. This is an adaptation to life in the sand. The fold of skin along the side of the body is very poorly developed.

Angolosaurus skoogi (Andersson, 1916)
Plated Sand Lizard

Distribution: Namib Desert from Namibia to southern Angola.
Description: A medium-sized lizard with a total length of 20 to 25 cm. Only in an extreme case will this lizard reach a length of 30 cm. The snout is spear-shaped. Head and body are cylindrical. The head is scaled and the nostrils lie between two nasal scales and the first labial scale. The ear opening is completely covered by a large scale. The smooth, small dorsal scales form 32 to 35 longitudinal rows and 62 to 65 transverse rows. The scales on the fold of skin along the sides of the body are granular. The scales on the underside of the body form eight longitudinal rows. The tail is slightly longer than the SVL and ends in a fine point. Males may be recognised by the well-developed pores on the thighs. These may be completely absent in females.

On the ivory-coloured, greyish white, or light yellowish brown upper side of the body there are numerous orange-coloured rectangular spots. The flanks are lighter in colour and often have large yellow flecks. Chin, throat, and chest are black, whilst the belly is white. Juveniles are a sand colour.
Habitat: Sand dunes.
Lifestyle, husbandry, and reproduction: These lizards usually live in colonies and hunt for beetles amongst the loose sand. They will also eat dry plant material, seeds, and grass. At the first sign of danger, they burrow into the sand at an amazing speed. They may remain submerged for more than 24 hours. In this way they also protect themselves from extremes in temperature. A dry vivarium with a ground area of 1 m^2 is adequate for a pair of these lizards. The substrate should be a deep layer of sand, whilst the other furnishings should be a pile of rocks, firmly cemented together. During the day, a spotlight should produce a temperature around 35 °C. At night, 18 to 29 °C is necessary. This species will eat large insects and small pink mice and will also lick on fruit. A rest period of 3 to 4 weeks without lighting and heating (at 18 to 20 °C) will stimulate subsequent mating. Mating usually takes place 1 or 2 weeks after the heating and lighting are again switched on.

Lizards

Cordylosaurus subtessellatus

Cordylosaurus Gray, 1865

Monotypic genus found in southern Africa. These lizards are very brightly coloured, and have well-developed limbs and prominent side flanks. The tail is extremely long.

Cordylosaurus subtessellatus (Smith, 1844)
Blue-Black Plated Sand Lizard

Distribution: Southern Angola through western Namibia, Namaqualand, and the western Cape.
Description: This lizard, which is normally 12 to 14 cm long, may even reach a length of up to 16 cm. The short head is covered with large scales. The nostrils are situated between two nasal scales and the first labial scale. The lower eyelids have a transparent scale. The smooth to heavily keeled dorsal scales form 15 longitudinal rows and 52 to 55 transverse rows. The scales on the underside form eight longitudinal rows. There are femoral pores. The tail may reach double the SVL. Along the back, this lizard is dark brown to black in colour, with prominent cream to yellow stripes. On the tail, these stripes change to blue. The underside is white.
Habitat: Rocky and stony areas with succulents.
Lifestyle, husbandry, and reproduction: This lizard is very fond of basking and will

usually lie on the sand with the legs stretched upwards. They hunt amongst the rocks for flies, crickets, and other insects. They may often be very closely approached before disappearing amongst the rocks at amazing speed. If they are molested, they will immediately cast off the tail which regenerates very quickly. Very little is known about their reproductive behaviour in the wild. Females usually lay only two eggs. This species should be kept in the same way as *Angolosaurus skoogi*.

Gerrhosaurus Wiegmann, 1828
Plated Lizards

Plated Lizards live in south and east Africa, in dry, often stony, semi-open regions. Six species are known, with total lengths between 40 and 90 cm. They have a powerful body which is only slightly compressed. In comparison to the body, the limbs appear quite delicate. The head and neck are barely set off from the body and the tail reaches exactly double the SVL. The scales, some of which are quite large, form regular longitudinal and transverse rows. The colouring is mainly brown tones, often with longitudinal stripes.

Gerrhosaurus flavigularis Wiegmann, 1828
Yellow Plated Lizard

Distribution: South and east Africa.
Description: *G. flavigularis* reaches a total length of up to 45 cm, of which two-thirds are the tail. On the brown ground colour of the body there are two pale yellow stripes, bordered in black, stretching from the neck to the tail joint. The back is covered by multikeeled scales, whilst the side folds have smaller scales. Males may be easily recognised by their prominent femoral pores, which have waxlike spikes several millimetres long.

In males, the throat is a yellowish colour whilst that of the females is a dirty-white tone. The ground colour of males is also less vivid than that of females, being more of a chocolate brown rather than a light yellowish brown.
Habitat: Between and beneath rocks and boulders on hilly and rocky terrain. Also in coastal areas where they usually excavate a hollow beneath a rock.
Lifestyle, husbandry, and reproduction: All Plated Lizards enjoy digging and are excellent swimmers and divers. It is not uncommon to find them living in termite mounds. In the wild they eat all manner of arthropods and will also take the young of small mammals and parts of some plants. This species is also very partial to sweet fruit. A clutch usually contains up to six calcareous, hard-shelled eggs.

A vivarium with a ground area of 100 × 60 cm is large enough for a pair of this species. Because these lizards like to burrow, a deep layer of sand, at least 10 cm deep, should be provided as substrate. Stacked flat rocks complete the furnishings. One of these flat rocks should be heated by means of a heating cable. Additional heat is provided by a spotlight. The rear wall of the vivarium can be easily made to resemble a cliff face by using polystyrene, plaster of Paris, and sand. During the day, temperatures of 25 to 28 °C are necessary, localised beneath a spotlight to 35 °C. At night the temperature should be reduced to 18 to 20 °C.

Because this species tends to obesity, they should be fed at most twice weekly. They will eat all manner of arthropods, pink mice, and fruit.

Kober (1990) saw his animals mating several times during November and December. The male pursued the female for a short time, biting her gently on the sides, on the neck, and finally on the throat until copulation resulted. This lasted between 5 and 10 minutes. Not long after mating, the female began to gain visibly in girth and at the end of December began to dig exploratory holes. Finally, three eggs measuring some 32 × 17 mm were laid. They were incubated at a daytime temperature of 28 °C and a night temperature of 22 °C. Almost 100% humidity is needed during the early part of the incubation period, otherwise the eggs will collapse. Later, atmospheric humidity of 80% is sufficient. During development, the eggs almost double in volume. One of the three eggs remained smaller than the others and hatched after 122 days. The remaining two larger eggs did not hatch until after 153 days. The young measured some 15 cm in length and fed on small crickets and house flies from the first day after hatching. Although they were essentially the same colour as the adults, the young had a number of white spots along the flanks and a line, composed of white spots, along the centre of the back. The young should always be reared separately from the adults and away from any other lizards.

Gerrhosaurus major Duméril, 1851
Sudan Plated Lizard, Brown Plated Lizard

Distribution: East and southeast Africa. The four known subspecies of Sudan Plated Lizard are very similar in appearance and are very difficult to distinguish. It is only the nominate form that is usually commercially available, since the remaining forms are only found in small parts of the distribution range. The systematic placing of the individual forms is still the subject of great debate amongst herpetologists since it is essentially only a matter of a variety of origins.

Description: The Sudan Plated Lizard reaches a maximum total length of up to 56 cm. Its dorsal scales are dark brown with a single keel. The throat is pink. The scales on the underside are in ten longitudinal rows. The underside of the tail is an ochre colour. In some subspecies there are fine black longitudinal lines.

Habitat: Dry savannahs and rocky regions.

Lifestyle, husbandry, and reproduction: This species lives in holes in the ground that it digs itself. It may also be found in termite mounds and rock crevices. They are diurnal and during the hottest part of the day will retreat to their burrows. John (1980) observed these animals in the wild. In the dry season, from January to March, the temperature averages around 33 °C. At the end of the rainy season, the temperature is around 25 °C. The atmospheric humidity increases after a rain shower. Al-

Gerrhosaurus major

though these animals do not live in groups or colonies, there is no great degree of aggression amongst them. Even in winter, the temperature does not fall below 20 °C. Examination of the excrement showed that the Agate Snail, which lives in the same areas, forms an important part of the diet of the Sudan Plated Lizard. Otherwise, they eat essentially the same food as *G. flavigularis*. In the vivarium they will eat crickets, small locusts, and other large insects as well as nestling mice, sweet fruit, canned cat food, and fruit yoghurt (Rese, 1981).

This lizard is easy to keep and is long-lived. With good husbandry it may live up to 15 years or more. Because of its size however, it needs a spacious vivarium with a ground area of at least 150 × 80 cm. A vivarium of this size is suitable for a pair or one male and two females. Sharp sand or fine gravel, at least 6 to 10 cm deep is a suitable substrate, because this species enjoys burrowing and digging. Climbing branches and some flat, stacked rocks complete the furnishings. If a medium-sized water container is provided, this will not only be used for drinking, but also for bathing. A water container from which the lizards may at least drink is however essential. Part of the substrate should be heated from below by means of a heating cable that must be firmly cemented-in so that it may not be dug out. From above, a spotlight should heat the vivarium to around 35 °C. At night the temperature may

Lizards

be allowed to fall to 20 to 22 °C, although even 18 °C will not cause any problems. A cooler rest period of around 3 weeks in July will stimulate both breeding and general physical condition. It is not wise to subject these lizards to constant temperatures. If sufficient hiding places are provided there will not be any aggression amongst these lizards. Breeding presents no problems. Females lay two or three eggs in a moist place. Incubation requires 110 to 145 days. On hatching, the young are around 10 cm long and should be kept under the same conditions as the adults, but must be reared separately. The young must also be given a regular and adequate supply of a vitamin/mineral preparation.

Gerrhosaurus nigrolineatus Hallowell, 1857
Striped Plated Lizard

Distribution: Eastern and northern Transvaal through Zimbabwe bordering onto northern Namibia to Gabon and the lower Congo.
Description: This 35 to 45 cm long species has a large, powerful head. Individual specimens may, in exceptional circumstances, reach a length of up to 56 cm. The eardrum shields are very narrow. The heavily keeled dorsal scales form 22 to 24 longitudinal rows and 54 to 64 transverse rows. The ventral plates are arranged in eight rows. In both sexes there are 16 to 18 femoral pores.

On the reddish brown back, these lizards have black-bordered yellow stripes. Frequently there are also additional yellow stripes. The flanks usually have irregular yellow flecks on a chestnut brown background. The underside is cream to yellowish white. Juveniles often have irregular transverse bands along the back.
Habitat: Open bushland, savannahs.
Lifestyle, husbandry, and reproduction: These lizards are often found in old termite mounds or meerkat burrows. At the slightest disturbance, these very timid lizards will flee into their refuges. They feed mainly on crickets, beetles, and snails. *G. nigrolineatus* should be kept under the conditions described for *G. major*. In the wild, females lay four to six eggs amongst piles of rotting vegetation. After an incubation period lasting 70 to 80 days, the young hatch measuring around 18 cm.

Gerrhosaurus validus Smith, 1849
Rock Plated Lizard

Distribution: Southern Africa northwards to Angola and Mozambique.
Description: These lizards are fully grown at a length of 60 to 75 cm. There are two subspecies. The nominate form has a dark brown to black upper side. The head and each of the dorsal scales has a yellow spot. From the head to the root of the tail there are two yellow side stripes. With increasing age, the stripes become paler and eventually only reach to the centre of the body. There are light perpendicular stripes on the flanks. The underside is light with some brown flecks. Juveniles are more intensively coloured (Matz, 1973).

Gerrhosaurus validus

Habitat: Rocky areas that provide numerous hiding places. Vegetation is usually restricted to dry grasses and Euphorbias (Switak, 1979, 1988).

Because of their size and temperament, the Rock Plated Lizards require the largest possible vivarium containing very stable furnishings. The substrate should be a 10 to 20 cm deep layer of river sand, part of which should be kept permanently moist. Flat, stacked rocks, firmly cemented together, will provide suitable hiding places. The vivarium may be planted with Aloes and Euphorbias, but these must be firmly embedded in the substrate. During the day, these lizards require a temperature of 22 to 26 °C, localised beneath a spotlight to at least 35 °C. At night, the temperature may drop to around 15 to 20 °C. Nestling mice, vegetables, lettuce, fruit, earthworms, insects and their larvae, and all other arthropods are all suitable foods for this species. They will also occasionally eat canned cat or dog food (Matz, 1973; Switak, 1988). A shallow drinking bowl is essential as is regular ultraviolet light.

In the wild, during the South African spring, females lay up to four eggs during the months of September to October. In the northern hemisphere the animals must first become acclimatised to the different seasons before they can be expected to breed. In South Africa, this species is frequently bred in outdoor enclosures. As

Lizards

proof that these animals may be acclimatised to the seasons of the northern hemisphere, in February a captive female laid three elongated eggs measuring 5 × 2.5 cm.

Tetradactylus Merrem, 1820
Whip Lizards

This small lizard is at home in South Africa. Some species have complete, though minute, limbs, whilst in other species, the forelegs are absent and only the relicts of hindlegs remain. The elongated body scales form transverse bands. There is a side fold. The relatively small head is covered by large scales. In African species, the prerostral scale is absent. The lower eyelid has scales. Femoral pores are only absent in *T. ellenbergi*. The tail is greatly elongated. Most of these lizards are very snakelike in appearance. Their hindlegs remain minute and only serve as anchors when the animal climbs in bushes or other vegetation. The tail is three times as long as the body and is used to provide propulsion. Whilst they are "swimming" through grass it is almost impossible to catch them, because they immediately discard the tail. The tail does however regenerate very quickly because without it these lizards are almost helpless.

Tetradactylus africanus (Gray, 1838)
African Whip Lizard

Distribution: Natal to the Southern Cape of South Africa. The subspecies *T. africanus fitzsimonsi* can be found on the southern and eastern Cape.
Description: Only rarely do these lizards reach a length of 33 cm. A more usual length is 25 to 30 cm. The forelegs and hindlegs remain minute and have only one toe. In the subspecies *T. a. fitzsimonsi* the forelegs are absent. The nostrils lie between two nasal scales and above the first labial scale. The olive-coloured back has dark brown stripes running along the centre of each row of scales. The head has irregular dark brown speckling. On the sides of the neck there are short black bars. The underside is a pale olive colour.
Habitat: Grassy mountain slopes.
Lifestyle, husbandry, and reproduction: This lizard is an excellent hunter of crickets and other insects on the grasslands where it lives. It is diurnal, retreating beneath rocks or clumps of grass at night.

A vivarium of 80 × 40 cm is required for a pair of these lizards. One-half of the base should be filled to a depth of 6 to 8 cm with sharp sand, the other half with a mixture of sand and loam for planting. Roots and firmly fixed piles of rock will be used for climbing and hiding places. During the day the temperature should be 24 to 29 °C, localised beneath a spotlight to 35 to 40 °C. At night a temperature of around 20 °C is sufficient. During the cooler part of the year, these lizards require a rest period at lower temperatures. This will help to bring them into breeding condition. In captivity, *T. africanus* will eat the usual foods. The vivarium should be sprayed daily.

Tetradactylus breyeri Roux, 1907
Breyer's Whip Lizard

Distribution: Southeast Transvaal and neighbouring Natal.
Description: This species usually reaches a length of 18 to 22 cm, rarely up to 24 cm. Each of the forelegs ends in two minute toes, whilst the hindlegs have a single toe. The three supraocular scales touch the frontal scale; the nostrils are situated between the two nasal scales. The dorsal scales are heavily keeled and form 12 to 14 longitudinal rows and 70 to 72 transverse rows. Only two, or rarely three, femoral pores are present. The olive brown back has dark stripes becoming paler towards the sides. There is dark speckling on the head, and on the neck there are short brown bars. The underside is pale olive.
Habitat: Grassy hillsides in mountainous regions.
Lifestyle and husbandry: As described for *Tetradactylus africanus*.

Tetradactylus seps (Linnaeus, 1758)
Dwarf Whip Lizard

Description: This small, long-tailed lizard rarely reaches its maximum length of 20 cm. It usually remains at around 13 to 15 cm. Its limbs are very small but perfectly formed. The lower eyelid is covered by scales. Along the back, the scales form 13 longitudinal rows. The tail is twice as long as the body. The body is olive brown to reddish brown with the flanks being somewhat paler. On the head there are dark brown spots. There are dark-edged flecks on the upper lip. The white neck has short dark bars. The underside is olive to blueish grey.
Habitat: Coastal woodland and mountain plateaux.
Lifestyle, husbandry, and reproduction: In the wild these lizards are only rarely found although they are quite common. They usually bask amongst dense vegetation, and when in danger "swim" extremely quickly using their long tail as a means of propulsion. Their main foods are crickets, beetles, and other insects. Husbandry is as for *T. africanus*. In the wild, females lay two or three large, oval, cream-coloured eggs beneath rotting wood or leaf litter.

Tetradactylus tetradactylus (Daudin, 1802)
Four-Toed Whip Lizard

Distribution: Southern and eastern Cape.
Description: This snakelike lizard reaches a length of 18 to 24 cm, rarely 29 cm, and is extremely slender. They have minute, four-toed forelegs and hindlegs. The keeled dorsal scales form 14 longitudinal rows and 59 to 62 transverse rows. There are four

Lizards 55

or five femoral pores on each side. The olive-coloured back has a stripe on each side. On the sides of the neck there are short black and white bars. The underside is pale olive.
Habitat: Grassland and "Fynbos" (heathland with a Mediterranean climate).
Lifestyle and husbandry: These lizards are astonishingly fast and spend the night below clumps of grass. They should be kept in vivaria such as those described for *T. africanus*.

Tracheloptychus Peters, 1854
Keeled Lizard

These lizards live in Madagascar on river banks in tropical forests. In build they are reminiscent of true lizards and reach a maximum length of 20 cm. Their side fold is limited to a small section of the neck. The dorsal scales are heavily keeled. The ear openings may be closed by large moveable scales.

Tracheloptychus madagascariensis Peters, 1854
Madagascan Keeled Lizard

Distribution: The extreme south and southwest of Madagascar.
Description: These lizards reach a SVL of up to 69 mm and the tail may be twice as long. The colouring varies from very light brown to blackish brown. Along the back there are three light longitudinal stripes of which the central stripe is often somewhat darker and broader, ending shortly behind the root of the tail. The two outside stripes continue down the tail. The dark dorsal bands, which are edged in a lighter colour, are usually unicoloured or interspaced with small spots. On the flanks there are feint, longitudinal bands comprised of light and dark spots.
Habitat: As described for *T. petersi*.
Lifestyle, husbandry, and reproduction: As described for *T. petersi*.

Tracheloptychus petersi Grandidier, 1869
Peters' Keeled Lizard

Distribution: South and southwest Madagascar.
Description: This lizard may reach a length of 76 mm, whilst the tail may be double that. Of the two species, *T. petersi* is the most brightly coloured. The red sides of the body begin behind the head and may be interspaced with light flecks. The red colouring runs down to the hind legs. The light flecks may run together to form longitudinal stripes. The reddish brown upper side is sharply delineated by a dark and light longitudinal stripe. The side of the head is blue, which can continue down the neck and sides of the body, but which gradually becomes paler.
Habitat: These lizards live in the hottest and driest part of the island amongst

Tracheloptychus madagascariensis

thorny vegetation. Even in the darkest shadows, the temperature can climb to 40 °C. In August it is at its coolest at 17 to 20 °C.

Lifestyle, husbandry, and reproduction: Meier (1988) first observed these animals in the wild amongst the sand dunes north of Tuléar. They live on dunes both with and without vegetation, together with *T. madagascariensis* where they also compete with them. At night, the lizards burrow themselves into the loose sand in the dunes. Around 1 hour before sunrise they cautiously push the head above the sand.

Keeled Lizards should be kept in a vivarium that has a deep layer of sand as substrate. Stacked rocks, a tree root, and spineless Euphorbias or other succulents may be used as decoration. The areas around the flower pots should always be kept moist. A drinking bowl is essential and the vivarium should be sprayed each morning. The lizards eat all manner of arthropods. They require temperatures between 25 and 30 °C, even higher beneath a spotlight. At night the temperature should fall to around 20 °C.

For breeding, the natural climatic conditions should be imitated as far as possible in that a warm period of 2 to 3 months is followed by a cooler phase at 15 to 20 °C, after which the temperature is again raised. The vivarium should be sprayed at least twice daily so that the atmospheric humidity is increased. During this time the lizards will mate, with the female producing eggs some 3 to 4 weeks later, usually near the flower pots and in a hole that she has excavated previously. The eggs should be transferred to an incubator since they dry out very quickly. With high humidity and temperatures of 26 to 30 °C the young will hatch after around 3 months. The hatchlings should be reared individually because they are very aggressive towards one another.

Zonosaurus Boulenger, 1887
Girdled Lizards

All species of Girdled Lizards live on Madagascar and nearby islands. In appearance they are somewhat reminiscent of the genus *Gerrhosaurus*. These powerful lizards have a very poorly set-off head, and, according to species, may reach lengths of 20 to 70 cm. The tail is almost double the SVL. The fold of skin on the sides is well developed and the dorsal scales are ribbed. In comparison to *Gerrhosaurus*, the ventral scales are very heavily overlapped in the manner of roof tiles.

Zonosaurus aeneus (Grandidier, 1972)
Bronze Girdled Lizard

Distribution: Madagascar, mainly in the east and on the island of Nosy Bé.
Description: Similar to *Z. madagascariensis* but with an average SVL of 6.6 cm. In this species, the stripes along the sides sometimes continue onto the tail.
Habitat: A very adaptable follower of civilisation.
Lifestyle, husbandry, and reproduction: See *Z. haraldmeieri*.

Zonosaurus haraldmeieri Brygoo & Böhme, 1985
Green Madagascar Girdled Lizard

Distribution: In northern Madagascar (Diego-Suarez and the surrounding area). Presumably also in many other places.
Description: The average SVL is around 14 cm. The upper side of this lizard is a yellowish green, whilst the flanks are dirty-white, silver-grey, or light brown. Back, tail, and limbs are covered in small black spots that form longitudinal and transverse stripes along the back. On the brown head there are also some small black spots. Males may be distinguished by their well developed femoral pores and thicker tail root. In comparison to the patterning of the adults, juveniles have a light stripe along each side of the back.
Habitat: Meier (1989) found two specimens on a dead tree trunk in the immediate vicinity of a house. Others were found below rocks. The location was somewhat cooler and more moist than usual in this part of Madagascar.
Lifestyle, husbandry, and reproduction: These Girdled Lizards should be kept in pairs in a vivarium with a ground area of 90 × 50 cm and 50 cm high. A suitable

Tracheloptychus petersi (above), Zonosaurus haraldmeieri (below)

substrate is a 5 to 10 cm layer of an earth/sand mixture, one-third of which should always be kept moist. The heating cable should be installed in such a way that places of varying temperature are provided. Better still is a heating pad below the vivarium, since these lizards will invariably dig up the heating cable. Some pieces of tree root will provide refuges and allow the animals to climb at will. A spotlight above the highest part of the tree root should provide a localised temperature of around 40 °C. For other parts of the vivarium, a temperature between 22 and 28 °C is adequate.

In captivity, this species will eat field crickets, house crickets, and snails. They will not, however, eat fruit although they may have fasted for some time.

These lizards enjoy basking for long periods, but may also remain submerged in the substrate for equally long periods. At night they will burrow into the substrate beneath the tree root.

After a rest period from September to mid-October at temperatures of 15 to 20 °C, the temperature should again be raised and the vivarium sprayed two or three times daily. Very soon afterwards, the animals will begin to mate. The female lays around 3 to 4 weeks later depositing four to six white eggs in a hole that she digs herself and later covers. The eggs are around 2.7 cm long and 1.4 to 1.6 cm wide. They have a tough calcareous shell. The eggs should be transfered to an incubator and incubated at a temperature of 26 to 30 °C with high atmospheric humidity. Under these conditions the young will hatch after 88 to 112 days measuring from 10.4 to 11.5 cm. At a constant temperature of 29 °C they should hatch after around 90 days. The young should be reared in small individual vivaria where they may be reared on small field and house crickets (Rogner, 1990).

Zonosaurus karsteni (Grandidier, 1869)
Karsten's Girdled Lizard

Distribution: Southwest Madagascar.
Description: These Girdled Lizards reach a total length of up to 40 cm. Their evenly coloured light brown back has a brown-edged yellow stripe along each side. On the flanks there is a band-shaped section which starts as a reddish brown and changes to a blueish colour. All three colour sections are interrupted by four rows of almost rectangular white flecks. Snout and chin areas are bright orange. Males may be easily recognised by the thicker tail root and well developed femoral pores.
Lifestyle, husbandry, and reproduction: At the slightest disturbance, this normally peaceful lizard will flee into its refuge at amazing speed. At night they burrow into the substrate leaving no trace of their existence. These Girdled Lizards therefore need a spacious vivarium and a deep layer of earth/sand mixture as substrate. Some pieces of tree root complete the furnishings and a drinking bowl is essential. In captivity, this species will eat house crickets, field crickets, small pieces of ground beef, and small pieces of apple.

Zonosaurus karsteni (above), Zonosaurus laticaudatus (below)

In the breeding season, from October to January, these lizards will often spend many hours basking and will regularly eat large amounts. Outside these periods they may only very rarely be seen, often remaining submerged in the substrate for many days. *Zonosaurus karsteni* may only be kept in pairs. Mating takes place in the usual way after which the female soon begins to increase in girth. The 30 × 18 mm eggs are pure white with calcareous shells and are buried deep in the substrate. Breustedt (1990) bedded the eggs in a dry mixture of sand and peat and placed them in an incubator. At constant temperatures of 28 to 29 °C and atmospheric humidity of 90 to 95% the eggs doubled in weight and increased in size by several millimetres during their development. After 93 to 95 days the young hatched measuring 17 to 18 mm. The female laid a second clutch around 4 weeks after the first. In Breustedt's case, the female even laid a third clutch of eggs 6 weeks after the second clutch. However, of the four eggs in this clutch, two proved to be infertile.

Zonosaurus laticaudatus (Grandidier, 1869)
Western Girdled Lizard

Distribution: Southern and northwestern Madagascar
Description: This lizard reaches an average SVL of 16.4 cm. The tail is twice as long. The light longitudinal lines become broader towards the neck where they disappear. They reappear behind the eyes, as is the case in other species, and reach down to the tip of the tail. In males, the head, sides of the neck and throat are bright red. The top of the head is somewhat darker than the neck and without spots. Between the light bands on the back, there are alternating dark and golden yellow scales. At the top of the back, they may form transverse stripes. On the sides and limbs there are small light flecks which may have a partial dark edge. The underside is more-or-less cream coloured. The throat in males may be varying shades of red.
Habitat: Has been seen in very dry as well as very moist places.
Lifestyle, husbandry, and reproduction: *Z. laticaudatus* should be kept in the same way as *Z. haraldmeieri* and *Z. karsteni* and is a very undemanding species that is very easy to care for.

Zonosaurus madagascariensis madagascariensis (Gray, 1831)

Distribution: Madagascar. At present from the east coast between Maroansetra and Tamatave. Also known from Nosy Bé. Individual specimens have been found in western Madagascar.
Description: This species reaches an average SVL of 12.7 cm with the maximum being 16.4 cm. It has the typical build of a Girdled Lizard. The yellow side stripes do not continue down the tail. These animals are brown to reddish brown in colour with numerous darker spots and flecks. The tail reaches double the SVL. On the sides and limbs there are small, lighter flecks which are partially bordered by a darker colour.

Zonosaurus m. madagascariensis (above), Zonosaurus maximus (below)

Habitat: A follower of civilisation, frequently seen in the vicinity of houses.
Lifestyle, husbandry, and reproduction: See *Z. haraldmeieri* and *Z. karsteni*.

Zonosaurus maximus Boulenger, 1896
Large Girdled Lizard

Distribution: Central east of Madagascar. Along the rivers Faraony, Matitana, Mananara, and Tolongoina.
Description: This lizard reaches a SVL of up to 24.6 cm with a tail length of 44.4 cm. One prominent characteristic is the laterally compressed tail. Juveniles are lighter than adults, but become darker with increasing age. This causes the black flecks along the back and the rows of yellow spots to disappear, especially in males. Later, the lower neck and flanks become grey with scattered blue specks. According to their origin, males have reddish or yellowish flanks which makes it easy to distinguish the sexes. The underside is greyish yellow and the throat sometimes has black flecks.
Habitat: Usually in the vicinity of water. Meier (1989) found that they were never discovered more than 2 metres from water.
Lifestyle, husbandry, and reproduction: In the wild, this species digs tunnels in which it lives. Like crocodiles, they live semi-aquatically. They are excellent swimmers and may remain submerged for several minutes. They are not so fast on land.

Because of their size, this species requires a spacious vivarium. In my collection, I house one male and two females in a vivarium measuring 140 × 80 × 140 cm, the substrate in which is a 15 cm layer of sand. A large tree root provides the necessary refuge. At night the lizards climb a large *Ficus benjamina* and spend the hours of darkness there. A spotlight above the tree root and a heating pad beneath the vivarium are switched on simultaneously with the lighting. These lizards rarely drink from the water bowl. If I place them in the water, they immediately leave it. Feeding is quite simple. They prefer canned cat and dog food, bananas and other fruit as well as crickets, locusts, and other large insects. In all other ways they are treated as *Z. haraldmeieri* and *Z. karsteni*. However, as opposed to these two species, *Z. maximus* is active throughout the year.

Zonosaurus ornatus (Gray, 1831)
Ornate Girdled Lizard

Distribution: Central eastern Madagascar. The eastern limit is the rain forest belt of the east coast.
Description: This species reaches a SVL of up to 13.2 cm with a tail length of 16.4 cm. Along the back there are two light golden-yellow longitudinal stripes, between which there is an additional light and two dark longitudinal lines. The head is dark brown and may have black flecks. The markings on the flanks are as similarly vari-

Zonosaurus quadrilineatus

able as in *Z. karsteni* and *Z. m. madagascariensis*. The underside is a whitish grey colour.
Habitat: Various regions and forests.
Lifestyle, husbandry, and reproduction: See *Z. haraldmeieri*.

Zonosaurus quadrilineatus (Grandidier, 1867)
Four-Striped Girdled Lizard

Distribution: Madagascar. In the region around Tuléar.
Description: In comparison to other species, these lizards are quite stout with only a short tail. They reach a SVL of 16.3 cm. The flat tail is half or one-third of the SVL. This shiny black lizard has four yellowish longitudinal lines, the centre two of which may run at an angle or end at the tail joint. Alternatively, they may run together and continue as a single line down the tail. The outer lines are straight and unite at the tip of the tail. The pattern on the sides forms a band of small light flecks. There are small light flecks on the forelegs and hindlegs. The underside is pink to creamish white.
Habitat: In the region around Tuléar the temperature in the middle of the year is around 23.8 °C. It can however climb to 39.8 °C and fall to 6.1 °C. With a rainfall of 750 to below 350 mm during the middle of the year, the southwest is the driest part of Madagascar, especially in the winter. In summer, very light rainfall may be followed by torrential downpours. The vegetation consists of Euphorbias, Aloes, Mimosas, and similar plants. This species is not a follower of civilisation.
Lifestyle, husbandry, and reproduction: To date, only a little is known about the lifestyle of this species. Meier (1988) watched a specimen eating red berries amongst

Zonosaurus rufipes

sand dunes. They should be kept in the same way as *Z. karsteni* with slightly higher temperatures. For breeding, an annual cycle of temperature and humidity variations is necessary.

Zonosaurus rufipes (Boettger, 1881)
Red-Footed Girdled Lizard

Distribution: Madagascar (endemic to Nosy Bé).
Description: These lizards reach a SVL of up to 8.8 cm whilst the tail may reach a length of up to 12.7 cm. They are easily mistaken for the young of other species. The more-or-less dark brown ground colour may have various tones between the back and flanks. This species also has two light longitudinal lines; they consist of two rows of small light rectangular spots. Each spot consists of a single black-edged scale. They alternate with black flecks on the sides of the tail, forming plainly visible longitudinal lines. The dark colour on the sides of the body has scattered, small, light spots. The colour of the underside ranges from grey to yellow or medium red.
Habitat: A forest dweller that is not a follower of civilisation.
Lifestyle, husbandry, and reproduction: See *Z. haraldmeieri*.

Zonosaurus trilineatus (Angel, 1939)
Three-Striped Girdled Lizard

Distribution: Madagascar. From Fort Dauphin to north of Ampany.
Description: This lizard reaches a SVL of up to 152 mm and in appearance is essen-

tially similar to *Z. quadrilineatus*, but with three lines instead of four. Juveniles have dark flecking.
Habitat: Similar to *Z. quadrilineatus*.
Lifestyle, husbandry, and reproduction: See *Z. quadrilineatus* and *Z. haraldmeieri*.

CORDYLIDS
Cordylids, Sungazers

With the exception of *Chamaesaura*, Cordylids have a very strongly compressed body. The head is well set-off from the body and from above appears triangular in shape. The scales are usually heavily keeled and spiny, especially at the root of the tail. An exception is the relatively smooth scalation of the *Platysaurus* species.

Chemaesaura Schneider, 1801

These snakelike lizards live in the hilly grasslands of southern Africa. They are extremely slender, have a pointed head, and frequently the tail reaches three times the SVL. According to species, they may reach a total length of 40 to 65 cm. The limbs are only rudimentary but may have five toes on each of the forelegs and hindlegs. In some species the forelegs are completely absent. The predominantly brown body is covered by large, heavily keeled scales. Along the back there are usually dark longitudinal stripes.

Chamaesaura anguina (Linnaeus, 1758)
Cape Snake Lizard

Distribution: Southern Africa northwards to Angola and Mozambique.
Description: *C. anguina* has four one- or two-toed limbs and reaches a length of up to 60 cm. The dorsal surface is chocolate brown. Along the centre of the back there is a narrow cream coloured longitudinal stripe. A wider stripe of the same colour stretches along the sides.

The diurnal *Chamaesaura* moves in a serpentine manner, does not burrow, and feeds on various arthropods. In their natural habitat they are subjected to considerable variations in temperature, therefore, in the vivarium there should be differences of at least 10 °C between the day and night temperatures. This species is ovoviviparous, the females giving birth to two to six, or rarely even up to ten, young.

Niekisch (1981) kept a specimen of this species in a vivarium measuring 50 × 25 × 23 cm covered with fine gauze. A 60-W incandescent lamp provided both lighting and heating. The substrate was a mixture of fine gravel and sand. Two pieces of boiled tree bark provided hiding places. In these hiding places the lizards would obviously feel secure, being able to have contact with the walls on all sides. These

Lizards

Chamaesaura anguina

snakelike lizards are mainly active during the morning and early afternoon, gliding around the vivarium in a serpentine manner. Niekisch fed his specimen on newly hatched grasshoppers, field crickets, house crickets, and spiders. The vivarium should be sprayed each morning so that the lizards may slake their thirst by licking up the resulting water droplets. At birth, young measure around 15 cm in length.

Cordylus Laurenti, 1768
Sungazers

The distribution range of this genus stretches from the southern part of Africa to East Africa. Within this range there are some 20 species living in dry, rocky areas. According to species, these lizards may reach lengths of up to 15 to 40 cm and have a large triangular head. The powerful, more-or-less compressed body is covered by large keeled scales. There are usually spiny scales on the back of the head and on the flanks. The tail is segmented and also covered in spiny scales. Males may be easily recognized by the well-developed femoral pores.

Cordylus cataphractus Boie, 1828
Armoured Sungazer

Distribution: Southwest Africa.
Description: Armoured Sungazers may reach a total length of 21 cm. These lizards are yellowish brown in colour with scattered dark spots. In comparison to *C. cordylus* which has 20 to 30 rows of scales, *C. cataphractus* has only 15 to 17 rows of scales

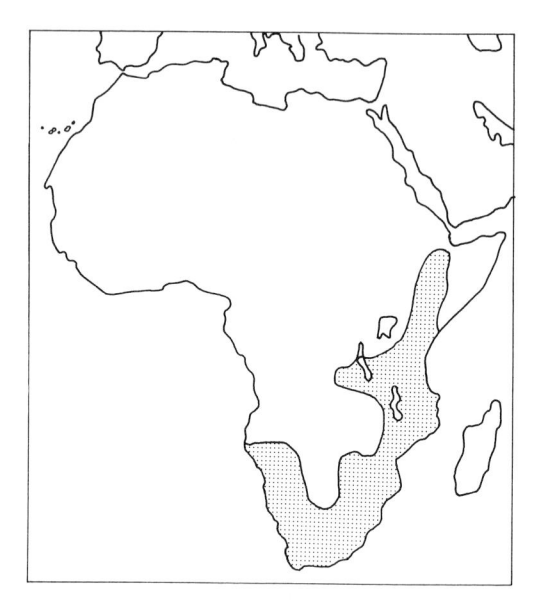

Distribution range of *Cordylus*

along the back. This species also has much stronger armouring. The neck and tail have especially spiny scales.

Habitat: Rocky, scree-covered hillsides with many narrow rock crevices. In winter light frosts and snow are possible.

Lifestyle, husbandry, and reproduction: When danger threatens, the sungazer will often retreat into a rock crevice where it will inflate its body and thus raise its spiny scales, preventing it from being removed. Their food consists of various arthropods, above all hard-shelled beetles. If the lizard is unable to reach its refuge, it will roll into a ball, biting its tail a few centimetres from the end, thus protecting its vulnerable underside. In even greater danger it may even discard the tail (Switak, 1987).

Most *Cordylus* species quickly become tame in captivity and are not aggressive towards one another. They are ovoviviparous, giving birth to between one and six young each year.

To accommodate this species comfortably, they require a number of hiding places and the deepest possible layer of sand as substrate. The air temperature should be around 30 °C whilst part of the base of the vivarium should be heated to 45 °C. It is essential that at night the temperature be reduced to 18 to 29 °C. It is equally important that a spotlight be provided for basking. A 4 to 8 week rest period at lower temperatures is also advisable to bring the lizards into breeding condition. These lizards hibernate in their natural habitat.

When keeping these lizards it is wise to have a breeding group of one male and

Cordylus cataphractus

two females. Flat rocks should be stacked in such a way that they provide adequate hiding places of different sizes. The rocks should be stacked in position before the substrate is introduced. This prevents the sungazers from burrowing beneath the rocks and causing them to collapse, which in turn could cause fatalities. The hiding places should be created in such a way that there is only space for one lizard. This prevents the lizards from injuring one another with their sharp spiny scales.

Sungazers require a great deal of sunshine and the entire vivarium should be brightly lit. The temperature below the spotlight should be around 40 °C. These lizards enjoy basking on the highest of the stacked rocks and it is above this point that the spotlight should be installed. These lizards should also be given ultraviolet light regularly—twice weekly for 15 to 20 minutes using an Osram Ultra-Vitalux™ lamp. This is especially beneficial for juveniles. These lizards will eat all manner of insects.

As a precautionary measure, gravid females should be placed in individual vivaria, because the young—usually one or two—will be attacked by the adults. When born, the young measure around 6.2 cm, which is around one quarter the length of the adults. Their spiny scales are still soft so that they do not damage the parturition channel. After having given birth, the female may be again replaced in the communal vivarium and the young reared in the "nursery suite." During the warmer months of summer it is advisable to keep these lizards in a gauze-covered outdoor enclosure. Unfiltered sunlight is extremely good for them and very difficult to substitute artificially.

Cordylus cordylus (Linnaeus, 1758)
Dwarf Sungazer

Distribution: Southern Africa northwards to southern Ethiopia.
Description: These lizards reach a maximum total length of 18 cm and vary in colour from black to orange-brown. Dwarf Sungazers are only slightly spiny.
Habitat: Dry, stony areas, less frequently on rocky hillsides. Also found below fallen trees and beneath tree roots.
Lifestyle, husbandry, and reproduction: See *C. cataphractus*. In the wild, each individual has a burrow in which it sleeps and into which it will flee in times of danger. Dwarf Sungazers also absorb moisture through the skin as a result of capillary action. Kutschera (1976) obtained a 17 cm specimen of this species on June 28, 1989. On November 29, at around 1500 h he found a newly born young measuring 6 cm in length. The young was still wet and soft and was lying close to the mother. The female appeared to be very weak and thirsty. The young was placed in a small vivarium. A second young appeared the following evening at 2030 h. Both young willingly accepted earthworms, grasshoppers, beetles, larvae, flies, moths, mealworms, and lean beef cut into fine strips. They occasionally lay in the water bowl. One of the young burrowed beneath a flat rock that collapsed and broke its neck. The second

was later rehoused with its mother; they even hibernated together between the stacked rocks, although the heating and lighting were still in operation. The juvenile was very aggressive towards other lizards.

Cordylus giganteus Smith, 1844
Giant Sungazer, South African Sungazer

Distribution: Southern Africa.
Description: The Giant Sungazer reaches a total length of up to 35 cm. Its tail does not break off as easily as that of other species. These lizards are brown in colour and have very long, spiny scales.
Habitat: Within their natural habitat there are fairly great changes in climate. During the summer months, from May to August, it is often extremely cold with minus temperatures and snowfalls. In summer however, it is hot and moist (Switak, 1980). According to Langerwerf, these lizards are mainly found in grassy areas on the high plains. During the course of the year there are storms on approximately 88 days. The average rainfall is 60 to 80 cm. On 60 to 90 days, the temperature may fall to the freezing point. It is only on 10 to 30 days each year that the temperature rises above 30 °C.
Lifestyle, husbandry, and reproduction: See *C. cataphractus*. In the wild, several adults will live together in underground burrows. Juveniles are only rarely found amongst adults. Sungazers often wander far away from their refuges and only seek them out when danger threatens. They frequently spend several hours with outstretched legs facing the sun. When the position of the sun alters, the lizards change their position accordingly.

Giant Sungazers are best kept in groups. These lizards require a large, spacious vivarium in which they may dig a burrow. According to Switak, there is "no lizard which is so easy to care for and which so quickly becomes tame. The Sungazer is long-lived and very easy to breed." Artificial hiding places are easily made from clay and should contain a deep layer of sand into which the lizards may burrow. During the day, this species requires a temperature of 25 to 27 °C, localised beneath a spotlight to 40 °C. At night, the temperature should fall to 11 to 17 °C (Paulduro et al., 1987).

These lizards will willingly accept grasshoppers, house crickets, field crickets, nestling mice, canned dog food, and occasionally fresh fruit. A bowl of fresh water should always be available. A winter rest period of 4 to 6 weeks, with a daytime temperature of 10 to 20 °C, falling at night to 8 to 10 °C, will help to bring the lizards into breeding condition. Females are gravid for around 4 months and usually give birth to two young. Vanderhaege (quoted by Matz, 1973) watched the birth of four young in March. After giving birth, the mother usually eats the embryonal sac. At birth, the young are around 12 cm in length, reaching some 18 cm around 8 months later.

Cordylus giganteus (above), Cordylus jonesii (below)

Lizards

Cordylus jonesii (Boulenger, 1891)
Jones' Dwarf Sungazer.

Distribution: Southern Africa; southern Mozambique, and southern Zimbabwe. Introduced into the eastern Kalahari in Botswana, Transvaal and Kimberley in the Cape Province through wood transport.
Description: These Dwarf Sungazers may reach a total length of 124 mm of which the SVL is 70 mm. The back of the head has six slightly protruding spiny scales. The scales on the back are heavily keeled. These animals are very variable in colour, being a dirty-grey, greyish brown, or reddish brown. Along the centre of the back there are often irregularly arranged dark brown or blue-grey spots which are sometimes arranged in symmetrical pairs. Juveniles often have a scattering of light spots.
Habitat: Dry savannahs, sometimes with large groups of trees, and semi-deserts. Here the lizards live beneath the loose bark of dead trees, in hollow trees or in holes in trees, often in the vicinity of human habitation or cultivated areas. In areas without trees, they seek refuge beneath rocks (Switak, 1980).
Lifestyly, husbandry, and reproduction: See *C. cataphractus*. This species is also easy to feed in captivity. It will eat grasshoppers, field crickets, house crickets, and other insects and their larvae. In the wild they consume enormous amounts of termites which helps them lay down a layer of fat. It is sufficient to feed these lizards every 2 or 3 days, and the vivarium need only be sprayed two or three times each week.

These Dwarf Sungazers require a large, spacious vivarium with sand as substrate. Tree roots and flat rocks will provide hiding places. During the day, the temperature should be between 30 and 38 °C, falling at night to 13 °C.

After a 4- to 8-week rest period at 10 to 18 °C, the first mating usually takes place a few days later. The male slowly approaches the female and if she is willing to mate will bite her on the neck region whilst pushing his tail beneath hers until the cloacal openings are adjacent to one another. Mating lasts for around 1 minute. The female is gravid for around 4 months before giving birth to two to four young. Whilst gravid, the female should be given adequate vitamins and minerals thus ensuring that the young are born without any deformities. The young vary in size according to how many each clutch contains. In Krabbe-Paulduro's case (1987), the young measured 72 mm at birth and weighed on average 2.2 g. Four of the young measured 65 mm and weighed 1.8 g. Dwarf Sungazers are sexually mature at around 3 years of age.

Cordylus jordani Peters, 1862
Jordan's Sungazer

Distribution: Central Namibia.
Description: This species only rarely reaches its maximum length of 28 cm. They are more usually from 20 to 24 cm in length. These animals resemble *C. polyzonus*, but are however larger and more powerful. The root of the tail consists of only one row

Cordylus macropholis

of scales. The body is a uniform olive brown whilst on the neck there are small black spots. Juveniles are the colour of leather and have a typical dark flecked pattern.
Habitat: Rocky hillsides.
Lifestyle, husbandry, and reproduction: See *C. jonesii*.

Cordylus macropholis Boulenger, 1910
Large-Scaled Sungazer

Distribution: Western Cape, coast of Yzerfontain to Kleinzee.
Description: A small sungazer, round in cross section and usually reaching a length of 10 to 13 cm, rarely 15 cm. The nasal scales touch and separate the internasal scales. At the back of the head there are six heavily keeled scales. The heavily keeled dorsal scales form 16 to 18 transverse rows and 14 to 18 longitudinal rows. The ventral scales are keeled and arranged in 10 rows. On the thighs there is a pair of enlarged anal pores and 10 to 12 femoral pores. The tail is in segments of large spiny scales. The back and flanks are grey to olive-grey with regular dark markings. The underside is pale grey.
Habitat: Sand dunes and coastal areas.
Lifestyle, husbandry, and reproduction: These very timid sungazers are usually

found beneath dead wood from Euphorbia bushes, on sand dunes, and below driftwood on beaches. Between April and May, females give birth to one to two young measuring 6 to 7 cm. Details of husbandry and reproduction are the same as for *Cordylus jonesii*.

Cordylus polyzonus Smith, 1838
Multi-Banded Sungazer

Distribution: South Africa.
Description: Usually between 20 and 25 cm in length. A charming sungazer with a compressed body. Its nasal scales are small and slightly tubercular. The lower eyelids have transparent scales. Along the back, the scales are very small and smooth, whilst on the flanks the scales are keeled. The scales on the back form 38 to 46 transverse rows. The root of the tail has two rows of spiny scales. In juveniles the yellowish brown to dark brown dorsal surface is checked with pale cream and the tail has dark brown bands. Some adults may retain this colouring (Karroo) or become dark brown to black (Central Cape to Kleinmaqualand). They may also be uniform olive, olive-brown with orange-red flanks (northwest of the Cape), or even blue-grey (southwest of the Cape). All populations have the characteristic black spots along the back.
Habitat: Very dry areas.
Lifestyle, husbandry, and reproduction: These lizards are one of the most common species in the central Karroo Region of the Cape. They are usually found in loose colonies amongst rocks on the lower reaches of rocky hillsides. During the hottest part of the day these lizards bask on rocks and catch beetles and grasshoppers. The winter is spent in tunnels that they dig themselves at the foot of rocky hillsides. These very alert sungazers flee into their refuges at the slightest sign of danger. In late summer the females usually give birth to two, sometimes three or four, young. In the vivarium they should be kept in the same way as *Cordylus jonesii*.

Cordylus rhodesianus Hewitt, 1933
Rhodesian Sungazer

Distribution: Eastern highlands of Zimbabwe bordering onto Mozambique.
Description: These 13 to 16 cm sungazers have a very compressed body and a triangular head with smooth scales. Their small scales, which towards the tail become keeled, form 25 to 29 transverse rows and 20 to 26 longitudinal rows. The scales on the flanks are keeled whilst those on the underside are smooth. Both sexes have five to eight femoral pores. The root of the tail has large spiny scales. The olive brown dorsal surface has irregular black flecks, occasionally pale flecks on the upper flanks. The underside is yellowish or greyish-white in colour.
Habitat: Grassy hillsides. Tröger (written communication) also found these lizards

in stony and rocky areas, in rock crevices, and beneath individual rocks. To be able to dig their tunnels beneath them, these lizards require rocks of only 20 × 10 cm.
Lifestyle, husbandry, and reproduction: Throughout their life in captivity, these sungazers remain extremely timid. When the keeper approaches they will usually retreat immediately to their refuges. Tröger keeps his animals in a vivarium measuring 100 × 60 × 50 cm with a substrate of a mixture of clay and sand. This has the advantages of extreme durability whilst keeping in place the stacked rocks amongst which the lizards find hiding places. In his vivarium, Tröger uses slabs of slate around 15 to 25 cm thick. Between the slabs of slate, Tröger places pieces of half-round cork bark and the clay/sand mixture. The lighting is provided from a 50-W HQL lamp which also provides radiant heat. During the course of the year the temperature varies between 20 and 28 °C. In one part of the vivarium a constant temperature of around 35 °C is maintained. During the day the humidity is around 50 to 60%, increasing to around 75% at night. The lizards were easily acclimatised to the climatic conditions prevalent in the northern hemisphere, In spring the vivarium is sprayed two or three times daily with tepid water, whilst in autumn, spraying is carried out every second evening. During summer, a drinking bowl is provided. In winter the lizards are given a 4-week resting period at a daytime temperature of 18 °C falling to 16 °C at night. During this period, neither the vivarium heating nor lighting is in operation.

Mating usually begins around 4 to 6 weeks after the rest period and lasts for 2 to 3 weeks. During the breeding period, males become more aggressive towards one another. The young are born during late summer, signifying that the females are gravid for around 6 to 7 months. These sungazers will eat all manner of insects which should be liberally coated with calcium and minerals before being presented. Vitamins and minerals are frequently added to the water used for spraying the vivarium. During spring and summer the lizards are fed thrice weekly and in autumn twice weekly. The newborn young are around one-third the length of the adults and should be reared in individual small vivaria. Rearing presents no problems and the young may be given a winter rest period in their first year.

Cordylus warreni Boulenger, 1908
Warren's Sungazer

Distribution: Southern Africa (several subspecies).
Description: These animals reach a total length of 20 to 30 cm. Their triangular head and body are strongly compressed. The dorsal scales are spiny and, according to subspecies, are arranged in 22 to 46 transverse rows. Both sexes have between 7 and 14 femoral pores. The tail is covered in large keeled spiny scales that are arranged in segments. Between each row of spiny scales there is a row of smaller, but also keeled, scales. These animals are dark brown to black in colour with yellow flecks and/or

Lizards

Cordylus warreni

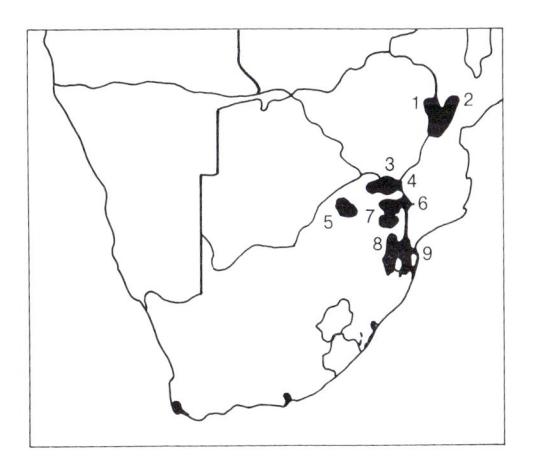

1. *C. w. regius*, 2. *C. w. mossambicanus*, 3. *C. w. depressus*, 4. *C. w. laevigatus*, 5. *C. w. breyeri*, 6. *C. w. vandami*, 7. *C. w. perkoensis*, 8. *C. w. barbertonensis*, 9. *C. w. warreni*

bands. In males, there are 3 to 4 rows of modified scales in front of the femoral pores. Males have a larger and wider head.

Habitat: Rocky areas with deep crevices on steep hillsides.

Lifestyle, husbandry, and reproduction: As in *Cordylus cataphractus*. In summer, from December to January, the temperatures soar from 30 to 40 °C, falling at night to around 18 °C. In winter, in June and July, the temperatures vary between 18 and 23 °C. During late afternoon they drop dramatically and at night may be as low as 8 °C. At higher altitudes, the temperatures may be even lower. These sungazers eat all manner of arthropods, frogs, lizards, and land snails. In the African high and late summer, the females give birth to two to six young. They do however require a spacious vivarium. Their excrement is always deposited in the same place, making the cleaning of the vivarium relatively easy.

Normally, these lizards mate immediately after the winter rest period. Females are gravid for 5 to 6 months. At birth, the young measure around 10 cm and weigh approximately 3 g. To be better able to keep a check on them it is advisable to rear the young in individual vivaria.

Platysaurus Smith, 1844
Flat Lizards

The natural range of the ten species of this genus is southern Africa. There they live in dry, rocky areas in crevices in granite, sandstone, or similar rock formations. In appearance they are somewhat reminiscent of true lizards of the family Lacertidae, and, according to species, may reach lengths of around 20 cm to well over 35 cm. The body is strongly compressed, the head large and equally compressed. The tail is around double the SVL. The body is usually brightly coloured and covered in small granular scales. Spiny scales are only present in the shoulder region. Their relationship to the family Cordylinae is shown by the large keeled scales on the tail. These are arranged in regular segments.

Platysaurus guttatus Smith, 1849
Small Flat Lizard

Distribution: South Africa.

Description: In contrast to many species of this genus, these lizards are not so brightly coloured. They may reach a length of 15 to 20 cm. The larger males are steel blue on the head and foreparts of the body. Other parts of the body are a greenish colour. Females remain a dull brown with a light yellowish stripe along the centre of the back. This stretches from the tip of the snout to the root of the tail. There is a further yellow stripe along each side of the back and light flecks along the flanks.

Habitat: In rock crevices and amongst boulders in dry areas. Often found amongst piles of granite on the savannahs.

Lifestyle, Husbandry, and reproduction: A dry, rocky vivarium measuring 90 × 30 ×

50 cm is adequate for a male and three females of this species. The rear wall of the vivarium should have many irregularities. This is easily created using plaster of Paris with pieces of rock or bark set into it. The lizards will spend much time here and will only descend to the ground to feed or drink. Flat rocks should be stacked in each of the rear corners. These will provide adequate hiding places and refuges for the lizards. Below a spotlight, the temperature should reach 45 °C, falling at night to 18 to 21 °C. One corner of the otherwise dry vivarium should be kept constantly moist and at around 25 °C. The vivarium should be sprayed daily. Food, in the form of all manner of insects, and their larvae should be given daily. A shallow drinking bowl should always be available.

Each year, the females lay one or two eggs measuring around 10 to 12 mm. In Freitag's case (1983), the females laid a total of five eggs, which, due to inattention, failed to develop. It is not wise to keep this species in an outdoor enclosure if they cannot be given protection from rain. During the year they should be given several rest periods at lower temperatures, but complete hibernation is not essential. Several subspecies are known.

Pseudocordylus Smith, 1838
False Sungazers

The *Pseudocordylus* species live in dry, rocky areas of South Africa. In the mountains they live at altitudes of up to 2,000 m above sea level. In appearance they are similar to the genus *Cordylus* but are even more compressed. The predominantly brown body is only very insignificantly spiny. The scales on the body do not have bony projections and its tail is much less spiny.

Pseudocordylus capensis Smith, 1838
False Cape Sungazer

Distribution: Individual populations from Cedarberg in the north through the Cape Rift Mountains to Kamanassieberg. Two subspecies: *P. c. roberti* (Cedarberg to Bokkeveld Mountains); *P. c. capensis* (Cape Rift Mountains).
Description: These graceful lizards usually reach a length of 18 to 22 cm, in exceptional cases 26 cm. Particularly noticeable are the long toes and slender tail. The flanks are completely covered in granular scales. The nasal scales touch one another and separate the rostral scale from the internasal scales. The tail, which is longer than the body, has spiny scales. Head and body are blue-black in colour. On the head there are yellow flecks which are also present on the body but less prominent. The underside is a uniform slate grey, somewhat paler in the centre. In some populations the throat is a pinkish red colour.
Habitat: Dry sandy areas with sparse, low vegetation.
Lifesyle, husbandry, and reproduction: These alert and lively lizards live in small, loose colonies between piles of rocks and boulders, preferably on vertical surfaces.

At the slightest hint of danger they flee to their refuges. Usually a male and female will be found in the same refuge. Their food is mainly insects. Females give birth during December and January, producing from one to three young. In the wild, these lizards feed mainly on bees and wasps. In captivity they should be kept in the same way as *Cordylus warreni*.

Pseudocordylus melanotus Smith, 1838
Black-Eared False Sungazer

Description: Escarpment Mountains from the Amatola Mountains to the Transvaal Drakensbergs.
Description: A medium-sized lizard, the scales on the flanks of which are separated by granular scales. The nasal scales touch one another and separate the rostral scale from the internasal scales. There are usually four supralabial scales. There may be up to 13 occipitals which may also even be absent. These lizards are extremely variable in colour. During the breeding season the colours of the males are especially bright and vivid. In females the back is grey to olive-brown with irregular pale flecks. The head, neck, and sides of the body are yellowish grey, yellow, or orange in colour. On the sides of the neck there are one or two dark flecks. The underside of these animals is whiter tending to light orange on the sides.

There are three subspecies. The nominate form has divided internasal scales and in females the femoral pores are absent. Males in breeding condition have 1 to 17 granular femoral scales and a wide, dark brown band on the back, sometimes with small pale flecks. The temporal area is reddish brown and on the throat there is a diffused blue-grey pattern.

The subspecies *P. m. subviridis* barely reaches a length of 25 cm. The internasal scale is undivided and the scales on the sides are larger than the spaces between them. Females also have femoral pores. Males in breeding condition have a black back with large, pale olive-grey flecks. Head and temporal region are olive-grey with a tinge of black. The flanks and sides of the tail are orange. The representatives of the subspecies *P. m. transvaalensis*, the taxonomy of which is not yet fully understood, have an undivided internasal scale. The scales on the sides are smaller than the spaces between them. Females also have femoral pores. Males in breeding condition are similar to males of the nominate form (Branch, 1988).
Habitat: Amongst rocks on mountain plateaux and in grassy landscapes.
Lifestyle, husbandry, and reproduction: These lizards live in large, loose colonies but only rarely is more than one lizard found on one rock. They eat small beetles and all manner of flying insects. During late summer the females give birth to one to four young after having been gravid for around 3 to 4 months. In the vivarium they may be kept in the same way as *Cordylus warreni*.

Langerwerf (written communication) kept three pairs in Alabama, U.S.A. in an

Pseudocordylus melanotus

outdoor enclosure of 72 m^2. They showed a considerable preference for the rocky part of the enclosure, but were also occasionally seen on the grassy parts. On August 10, 1991, six young were found in the enclosure. They were fully grown after only 13 months.

On July 20, 1992, a further two young were found. These were followed by another five young on July 21. The young are similar to the adults in all respects. It is Langerwerf's opinion that adult females may give birth to six or seven young, not a maximum of only four as previously thought. The young in his case, however, could have been produced by two females giving birth at the same time. Rearing the young presented no problems. In the area where Langerwerf had his enclosure, the summer months are very hot and moist with frequent storms. From May until the end of September, the temperature is between 28 and 35 °C. In winter, from November to March, night frosts are possible and snow falls every 2 or 3 years. On the coldest nights, the temperature may fall as low as 10 °C. The lizards hibernated in deep tunnels in the outdoor enclosure as did the *Cordylus giganteus* that live with them.

Even at outdoor temperatures of only 5 to 10 °C *Pseudocordylus melanotus* will emerge to bask.

Pseudocordylus microlepidotus (Cuvier, 1829)
Small-Scaled False Sungazer

Distribution: Southern Africa.
Description: With a total length of 32 cm, maximum 35 cm, these lizards are the largest of their genus and are very variable in colour. Their back is light yellow to blackish brown. The flanks are yellow or orange whilst the underside is cream. The flanks often have six light yellow triangles, whilst those on the back run together to form narrow yellow stripes. The tail and insides of the legs are also light yellow. The underside is dirty-grey, the throat black. On the tail there are only a few poorly defined spiny scales. The tail breaks off much more easily than in the true sungazers of the genus *Cordylus*. In Basutoland and Natal the animals are larger than those from the Cape Peninsula. They are also lighter in colour. Animals from Swaziland have a light red underside and are much flatter in build.
Habitat: Between rocks in the mountains. Animals from the Orange Free State and Eastern Transvaal are subjected to much greater heat than those from the mountains. The mountains are partially shrouded in mist, nevertheless animals from this region require an enormous amount of sunshine (Switak, 1980).
Lifestyle, husbandry, and reproduction: The diurnally active lizards require a vivarium with a large number of stacked rocks. Hollow building bricks are ideal for this purpose. These are very timid lizards and will never become as tame as *Cordylus* species. They will readily eat insects, newborn mice, and small amounts of canned dog food. They will occasionally also eat fruit and vegetables. Animals will also be found that are complete individuals having very distinct food preferences. A shallow bowl of fresh drinking water should always be available in the vivarium. These lizards are ovoviviparous, the female giving birth to three to six young (Matz, 1973). *P. microlepidotus* may be kept under the same conditions as *Cordylus warreni*.

LACERTIDS
Family Lacertidae

Lacertids inhabit an enormous distribution range which stretches from the subtropical and temperate regions of the Palearctics and Africa to southeast Asia. The family contains around 20 genera with over 200 species. These lizards occur in the most divergent habitats: deserts, steppes, light forests, rocky areas, and less frequently in tropical rain forests. Nearly all species have several things in common: a slender build, relatively long tail, and well-developed limbs. The top of the head has enlarged scales whilst the large ventral scales are usually arranged in longitudinal and transverse rows. Even when only very insignificant, there is a transverse fold of skin between the scales on the throat and the chest of most species. Enlarged scales with an unattached rear edge form a "collar." Lacertids almost always have femoral pores, especially males.

The tail is especially fragile. When threatened it may be cast off because from the sixth vertebra there are distinct breaking points. Regenerated tails never attain the original length. During the breeding season the males of many species develop extremely bright colours. The colouring of the juveniles is also usually much brighter than that of the adults. All species are diurnal and most require a great deal of sunshine. Because males will not tolerate one another and in the wild usually defend a well-defined territory, in captivity it is best to keep these species in pairs or in breeding groups of one male and two or three females.

During the breeding season, males follow females relentlessly. Only very primitive genera such as *Gallotia* and *Psammodromus* still inflict the bite to the neck which is common amongst other lizards. Apart from *Lacerta vivipara* and some *Eremias*, species that are ovoviviparous, all Lacertidae lay eggs.

The development of the eggs actually starts in the uterus. The female however only lays the eggs when the embryo has developed around fifty original segments. At that point, the embryo is still able to move and, should the position of the egg be changed, is able to turn upright again. Soon after the egg is laid however, this is not possible and the embryo suffocates when the position of the egg is changed. Even one day after being laid, a quarter of a rotation of the egg is sufficient to cause the death of the embryo. Eggs must therefore be transferred to the incubator in exactly the same position as they were found and should be incubated at 27 to 39 °C. During the course of their development the eggs sometimes increase considerably in size and weight.

In the wild, the eggs are usually subjected to variations in temperature and only rarely experience the almost constant temperatures found in an incubator. Breeders who incubate eggs at varying day and night temperatures usually achieve better results.

Acanthodactylus Wiegmann, 1834
Fringe-Fingered Lizards

The distribution range of Fringe-Fingered Lizards includes North America and the Near East. *Acanthodactylus erythrurus* is the sole European member of this genus and can be found in southwest Europe. A total of around 12 species inhabit sandy or rocky areas with sparse vegetation. According to species, Fringe-Fingered Lizards reach a total length of between 15 and 20 cm. They all have a well-proportioned body, a relatively pointed head and a thick tail root. On the rear half of the body the dorsal scales are significantly larger than those on the front half and are slightly pantile. These are long comb or fringe scales on the toes. The Fringe-Fingered Lizards are similar to the genus *Eremias*.

Acanthodactylus boskianus (Daudin, 1802)
Bosk's Fringe-Fingered Lizard

Distribution: North Africa to the Near East in the Mediterranean Region.
Description: These lizards reach a total length of up to 20 cm. Their yellow to orange-brown upper side often has seven white longitudinal stripes and light to dark brown flecks. The underside is white to yellow. Males may be easily recognised by the thick tail root and well-developed femoral pores.
Habitat: Sandy semi-deserts and deserts with only sparse vegetation. Also frequently found near human habitation.
Lifestyle, husbandry, and reproduction: Fringe-Fingers are best housed in a dry vivarium. During the day, the temperature should be between 25 and 36 °C, falling at night to 15 to 22 °C. Localised heating from a heating pad and a spotlight are absolutely vital. These lizards eat insects and spiders but will also take fine strips of beef and beef heart. A suitable substrate is an 8 cm layer of sand, one corner of which should contain plants and be kept constantly moist. It is here that the females will lay eggs. Succulents are particularly suitable plants for this type of vivarium. Stacked flat rocks provide adequate hiding places and refuges. During the winter these lizards should be kept at 10 °C after which they will start breeding.

After hibernation the male approaches the female with inflated throat and the neck bent downwards. He licks the sides of the female, carefully checking the root of the tail and the cloaca. Females that are willing to mate will lie with the body pressed to the ground and the root of the tail slightly raised. By nudging along her flanks with his snout, the male excites the female before biting her firmly on one side of the root of the tail. Copulation may last from 1 to 4 minutes (Zimmermann, 1983).

Around 2 weeks later the female explores the entire vivarium carefully checking the ground. At the place where the ground is moist, the female digs a hollow around 8 cm wide where she may lay up to seven eggs, which are then meticulously covered using both the body and the snout. The next mating may take place only a few days after the eggs have been laid.

The parchment-like eggs are around 1.6 cm long and 0.8 cm wide. They should be placed in fine moist sand or vermiculite and incubated at a temperature of around 28 °C. The yellow and black striped young hatch after 89 to 100 days measuring approximately 8.5 cm in length. The young should be reared individually in small vivaria and should be fed on small arthropods. They should not be hibernated during their first year and will be sexually mature after around 18 months.

Acanthodactylus erythrurus (Schinz, 1833)
European Fringe-Fingered Lizard

Distribution: Iberian Peninsula and northwest Africa.
Description: The European Fringe-Fingered Lizard reaches a SVL of 7.5 cm whilst the tail may be twice as long. Adults have a brownish, grey, or ochre-coloured upper side with up to ten light longitudinal stripes between which there are often rows of white, yellowish, or grey to black flecks. On the legs there are round white spots. During the breeding season the male develops round yellowish flecks along the flanks. The colouration is extremely variable. In eastern Spain there are Fringe-Fingers with greatly reduced markings. Juveniles have vivid black and white longitudinal stripes on the body, whilst the tail and thighs are bright red.
Habitat: Open landscape. These creatures prefer sandy ground, especially along the coast, but may also be found on rocky terrain with sparse vegetation.
Lifestyle, husbandry, and reproduction: See *A. boskianus*. This species only becomes sexually mature at around 3 years of age, and at that age a female will lay only one clutch of eggs. In subsequent years two clutches are more usual, each clutch containing four or five eggs measuring around 15×8 mm. In the wild, eggs develop in 70 to 75 days until the young hatch measuring 6 to 7 cm in length. If kept at a constant temperature of 30 °C the eggs hatch after only 35 to 40 days.

Acanthodactylus pardalis (Lichtenstein, 1823)
Speckled Fringe-Fingered Lizard

Distribution: Northeast Libya, northern Egypt, and Israel.
Description: A Fringe-Fingered Lizard with 12 to 14 rows of ventral scales and three rows of scales around each finger. These powerful 12 to 15 cm lizards have conspicuous yellow to light brown flecks on the dorsal surface. This species is bound to a particular habitat. They are however found on parched grassland and overgrown sand dunes.
Lifestyle, husbandry, and reproduction: Essentially the same as all other Fringe-Fingerd Lizards. A vivarium with a ground area of 60×40 cm is large enough for a pair. The substrate should be a 10 cm layer of sand. Rocks and pieces of bark will provide refuges. A spotlight should provide a localised temperature of 38 °C, elsewhere a temperature of around 20 °C is adequate. If an additional means of heating is available, these lizards may be kept in a covered outdoor enclosure during the summer months. A winter rest period of 1 to 2 months at a temperature of 4 to 5 °C will help to bring the lizards into breeding condition. Mating usually starts around 4 weeks after the winter rest period. When kept under optimum conditions the females will lay from three to seven clutches of eggs at intervals of around 20 days. Each clutch consists of two to five eggs from which, at an incubation temperature of

28 to 30 °C, the young will hatch after 55 to 65 days. They require a regular supply of vitamins and minerals and should be reared individually.

Algyroides Bibron & Bory, 1833
Keeled Lizards

The representatives of the genus *Algyroides* are close relatives of the genus *Lacerta*. They live in semi-shaded, slightly moist areas and barely reach a length of 15 cm. These slender lizards may be easily recognised by their large keeled dorsal scales. Along the flanks the scales are small and smooth. These lizards are usually very plain in colour. The upper side of the body is brown and the underside somewhat lighter.

Algyroides fitzingeri (Wiegmann, 1834)
Pygmy Algyroides, Sardinian Algyroides

Distribution: Corsica, Sardinia, and the island of La Maddalena off the north coast of Sardinia.
Description: With a maximum SVL of 4 cm and a tail measuring 8 to 9 cm, this is the smallest of the Keeled Lizards. Because of the equally large dorsal and side scales that are heavily keeled diagonally and come to a point this lizard is easily distinguished from others occurring in the same area. On the back and flanks the colouring varies from brown tones to a blackish olive. Irregularly arranged black spots are often found on the back. These may occasionally form a central stripe. On the underside these lizards may be blue, grey, yellow, or orange.
Habitat: Sunny, rocky shores but also in shaded gorges in the mountains (up to an altitude of 1,500 m) and in the vicinity of water.
Lifestyle, husbandry, and reproduction: See *A. marchi* and *A. moreoticus*.

Algyroides marchi Valverde, 1958
Spanish Algyroides, Spanish Keeled Lizard

Distribution: Very small populations on the mountains of Cazoria, Segura, and Alcaraz in southeastern Spain.
Description: These keeled lizards reach a total length of around 15 cm with a SVL of approximately 5 cm. Along the back there are small black flecks on a coffee-coloured background. Along the flanks there is a black or grey, light or dark stripe, the upper side of which has a brilliant edge. The underside is bright yellow. The throat and chin of males is white to blue-grey.
Habitat: Amongst scree and in clearings in pine forests in the vicinity of small mountain streams or ponds at altitudes between 700 and 1,600 m above sea level.
Lifestyle, husbandry, and reproduction: These small keeled lizards may be found on

Algyroides moreoticus

stone walls, on rock piles, and on bushes, especially in the vicinity of water. In dried-out stream beds they are normally found near residual pools of water (Eikhorst, 1979).

A vivarium of 60 × 30 × 35 cm is suitable for a small breeding group. Sand is a suitable substrate. Several pieces of bark stacked on one another will provide refuges. Moss-covered rocks, branches, or pieces of bark complete the furnishings. The moss should be sprayed daily to increase the atmospheric humidity. The lizards may approach a spotlight to bask by climbing a branch that has been installed at a suitable angle and height. The localised temperature below this spotlight should be around 40 °C. In other parts of the vivarium, temperatures of 22 to 25 °C are adequate. These lizards will eat all manner of small arthropods.

After a 10-week rest period at 4 to 6 °C the lizards will begin to mate. To hold her in position, the male bites the female on the flank, immediately in front of the hind leg. The actual mating lasts between 10 and 20 minutes. Three to four weeks later the female usually lays three eggs, measuring on average 11.2 × 6.4 mm. After a period of 21 to 24 days a second clutch follows, sometimes even a third. Eggs are usually laid just below the surface in moist sand. At temperatures between 26 and 30 °C the young hatch after 34 to 38 days measuring on average 7.2 cm.

Algyroides moreoticus Bibron & Bory, 1833
Ionian Algyroides

Distribution: Greece, Peloponnes, and on the southern Ionian Islands of Kephallenia, Ithaca, and Zakynthos.

Description: These keeled lizards reach a total length of around 15 cm. The upper side of the male is a dark to reddish brown with dark to black spots, with light spots along the flanks. Between the light spots there is a light stripe along the side. On the white to yellowish underside there are often black spots. Females are usually a uniform brown.
Habitat: Cliffs, old stone walls, and amongst scree. On hedges and bushes, and in forests in the vicinity of streams.
Lifestyle, husbandry, and reproduction: See *Algyroides marchi*. It is obvious that in the wild these lizards prefer north-facing hillsides and do not require a great deal of heat and sunshine. In den Bosch (1983) kept a pair in a vivarium measuring 60 × 40 × 40 cm with a substrate of peat and deciduous leaves upon which he placed some rocks and pieces of bark as refuges. The temperature was around 20 °C. A weak spotlight was also installed. This species eats field crickets, house crickets, and meadow sweepings. In the case of In den Bosch the female laid five to seven clutches of eggs each year, each clutch consisting of one to four eggs.

At a temperature of 29 °C and high atmospheric humidity the young hatched after 36 days. They grew very quickly and reached sexual maturity within 1 year.

Algyroides nigropunctatus (Duméril & Bibron, 1839)
Greek Algyroides, Greek Keeled Lizard

Distribution: From extreme northeast Italy to northwestern Greece including the Ionian Islands.
Description: With a total length of around 21 cm this is the largest of the keeled lizards. In comparison to *Algyroides moreoticus*, the heavily keeled dorsal scales appear much larger than the scales on the flanks. These lizards are brown, grey-brown, olive-green to black in colour. On the ground colour there are often irregular black spots that may sometimes be arranged in rows. Sexually mature males may be recognised, above all in the breeding season, by the vivid blue to violet throat and underside of the head. The underside of the body and tail, the insides of the limbs, and sometimes the flanks are orange to brick red. Females have a yellowish to greenish-white underside.
Habitat: Along coastal regions in vegetation-rich areas. Usually on the banks of streams, on hedges and bushes, on stone walls, cliffs, fences, refuse dumps, old olive trees, and frequently also in the vicinity of human habitation.
Lifestyle, husbandry, and reproduction: As in other *Algyroides* species. Adult males possess and vigorously defend a large territory which, during the breeding season, they usually share with one female. The breeding season begins at the end of April after a winter rest period. At the end of May the female lays two or three eggs. A second clutch sometimes follows in summer. During the hottest part of the year, these lizards lead a very secretive lifestyle.

Lizards

Eremias Wiegmann, 1834
Racerunners

The largest distribution range of this genus stretches from Africa through Europe to the Near East and Manchuria. The 22 species live in semi-deserts and deserts. In appearance they are somewhat reminiscent of *Acanthodactylus* but are distinguished from them by several characteristics. The nostrils are on a raised lump and have no contact with the supralabial scales. The dorsal scales are small and mostly finely granular. The rows of scales on the underside run together at an angle to meet in the centre. There are fringelike projections on each of the toes. The occipital scale is absent. In addition, some species may be recognised by a trace of transparent eyelids.

Eremias pleskei Nikolsky, 1905
Transcaucasian Racerunner

Distribution: Southern Armenia and adjacent regions of northeast Turkey and northwest Iran.

Description: This species reaches a length of 16 cm and has a grey to blackish brown dorsal surface. There are three whitish longitudinal stripes along each side of the body. These are especially pronounced in juveniles. There are large white flecks at the top of the legs. The rear of the femur and underside of the tail are yellow. In comparison to *E. strauchi* and *E. velox caucasia* this species is more graceful and has a longer tail.

Habitat: Steppes and dry barren land. Also in watered vineyards up to an altitude of 1,700 m.

Lifestyle, husbandry, and reproduction: As for *Acanthodactylus boskianus*. As ground dwellers, these lizards enjoy digging and spend the hottest hours of the day and nights in burrows that they dig themselves. They feed on all manner of arthropods.

Racerunners have a very short lifespan. The smaller species rarely live longer than 2 or 3 years, but reach sexual maturity at end of their first year. A clutch usually consists of two to six eggs. In the wild the young hatch after an incubation period of 2 to 3 months.

Liesack (1984) kept two males and six females with other species in a vivarium measuring 120 × 30 × 40 cm containing a substrate of fine hay some 4 cm deep. A large flat rock, a tree root, and several pieces of bark provided hiding places. Also installed in the vivarium was a plastic container of sand measuring 28 × 12 cm. This was invariably used for egg laying. A shallow drinking bowl was placed on the sand. Racerunners prefer however, to drink from water sprayed into the vivarium during the evening.

At some parts of the floor of the vivarium the temperature measured around 35

Eremias strauchi

to 40 °C, and at others it was even higher. The air temperature was between 30 and 35 °C during the day, falling at night to 20 to 25 °C

On April 6, 1981, a female laid three eggs in the moist sand container. At temperatures between 25 and 32 °C, one young hatched after 41 days measuring 6.8 cm. It was fed on small waxmoth larvae and aphids but only lived for 14 days. A very varied diet is required to successfully rear the young of this species. Ants form an important part of their diet in the wild.

Eremias strauchi Kessler, 1876
Strauch's Racerunner

Distribution: Southern Armenia, Nikitschevan, southern Azerbaijan; also eastern Turkey, northern Iran, and Turkemenia.
Description: With a total length of up to 20 cm this lizard is somewhat larger than E. pleskei. The uniform grey-brown centre of the back is bordered by two longitudinal rows of white spots which in turn are surrounded by black spots that may sometimes run together to form a dark dorsal stripe. Along each flank there is a row of greenish spots. The underside is white without any markings. The upper side of juveniles is especially colourful: dark brown with white stripes and rows of white spots, whilst the thighs and underside of the tail are bright yellow.
Habitat: Dry steppes, fallow land, and semi-deserts with sparse vegetation.
Lifestyle, husbandry, and reproduction: As described for E. pleskei. In the wild this

lizard eats mainly ants but will also eat other insects and spiders. They hibernate from late October until March, which in the vivarium is achieved by reducing the temperature to 4 to 6 °C. Mating begins in April, after which, during June, July and August, females lay one to three clutches of eggs. Each clutch may contain three to seven eggs. Eggs may even be produced towards the end of August. Because the life expectancy of this species is very short, the young are sexually mature after their first hibernation (Liesack, 1984).

Eremias velox caucasia Lantz, 1928
Caucasian Racerunner

Distribution: Transcaucasia, Dagestan and the Kalmukisch region.
Description: Along each side of the light greyish brown back, there is a light stripe below which there are two rows of black-bordered white spots. The lower spots are a greenish colour whilst the belly is white to yellow. In juveniles, the back of the thighs and the underside of the tail are red. Males may reach a length of 17.5 cm, whilst females remain smaller and paler in colour.
Habitat: Dry, rocky steppes, and on the dry Vermut Steppes.
Lifestyle, husbandry, and reproduction: Liesack (1981) kept two pairs together in a vivarium furnished as for *E. pleskei*. From late December until mid-March the lizards hibernated in a sand-filled glass container. One part of the ground area was dry, the other slightly moist. Although mating was not seen, one of the females laid four eggs on June 1 and a further two on July 12. These were deposited in the sand-filled container but incubation was not successful. On April 23, the female laid a further four eggs. Three of these hatched after 50 days, producing young measuring some 7 cm each. The fourth egg was accidentally damaged during the incubation period The young died after 5 to 7 days, suffering from convulsions that may have been prevented by giving vitamins, calcium, or ultraviolet light. Perhaps it is essential to feed the young on ants, but it is more likely that the convulsions occurred due to a lack of vitamin B in the female.

Gallotia Boulenger, 1916
Canary Island Lizards

During the past 15 years, great attention has been paid to these lizards which live exclusively on the Canary Islands. Earlier these species were assigned to the genus *Lacerta* but they differ from Lacertids in several respects, one of these being that *Gallotia* males do not bite the females before mating. These very dark coloured lizards emit squeaking sounds. Their ground colour is an adaptation to the dark backgrounds in their natural habitat. Males have only a few blue or greenish markings whilst females retain their juvenile stripes for some time.

Gallotia atlantica (Peters & Doria, 1882)
Atlantic Canary Lizard

Distribution: Lanzarote, Fuerteventura and some small neighbouring islands, Gran Canaria around Arinaga (introduced).

Description: This is the smallest member of the genus *Gallotia*, with the males reaching a total length of 28.5 cm. At 19.5 cm the females remain considerably smaller. In contrast to other *Gallotia* species, these animals have very large and significantly keeled scales. Their very variable colouring and markings often depend upon the habitat in which they live. Specimens that live on sandy ground are light brown with a very pale pattern, whilst those living on lava are a blackish brown. In the second instance the patterning is swamped by the darker colour, but all transitory patterns are possible.

On the upper side, both sexes are grey to olive-brown in colour. The markings consist of light and dark longitudinal stripes or a hint of barring. Along the flanks there are more-or-less light ocelations. Females are usually more uniform in colour with less prominent longitudinal stripes. In newly sloughed juveniles there is a very bright yellowish white stripe along each flank and two along the centre of the dark brown back. There are at least two known subspecies.

Habitat: These unassuming lizards inhabit almost all areas. Piles of rocks and stone walls provide refuges. There must however be some ground vegetation since it is here that the lizards hunt for food.

Lifestyle, husbandry, and reproduction: Canary Island Lizards emit a chirping or squeaking sound and will eat all manner of arthropods. For adults, flowers, leaves, and fruit form part of their diet. In captivity, this species appears to enjoy fruit yoghurt, banana, and tomato. Unfortunately throughout their life, Canary Island Lizards remain extremely timid. All species require very warm conditions. The temperature in the vivarium should be between 22 and 25 °C, localised beneath a spotlight to between 35 and 40 °C. A 4- to 6-week winter rest period will not only stimulate breeding but will also improve their well-being. At night the temperature should be reduced to 20 °C.

Although this species may not be exported without permission from the Spanish authorities, several herpetologists already have them in their collections and they are not particularly difficult to obtain. This species is frequently bred in captivity.

A pair requires a vivarium of at least 60 × 30 cm. A 4 to 5 cm layer of sand provides a suitable substrate. Near the shallow drinking bowl, the sand should be kept slightly moist. Several flat, stacked rocks will provide refuges. Fresh drinking water should always be available.

During the breeding season, the male approaches the female very quickly with downward-pointing snout and inflated throat, at the same time constantly bobbing his head. He then seizes the female by a fold of skin on the neck and completes the mating which may last up to 3 minutes. Females usually lay two clutches of eggs each year, each clutch consisting of two to six eggs measuring 10 to 15 mm in length and 7 to 8.4 mm in width. Newly hatched young have a SVL of 24.5 mm and a tail measuring around 43 mm.

Gallotia galloti galloti (Oudart, 1839)
Canary Island Lizard

Distribution: Central and southern Teneriffe. Intergrades with the northern subspecies *Gallotia galloti eisentrauti* may be found in the Teno Mountains in the northwest and around Esperanza.

Description: Males may reach a total length of 37 cm whereby 12.5 cm is SVL. Females may reach a length of 34.5 cm. of which the SVL is 11.5 cm. At the front of the body the more sturdily built males are a very deep black whilst the remainder of the body, tail, and hindlegs are grey-brown. Some males are completely black. Along the back there may occasionally be an almost indistinguishable pattern of several irregular transverse bands or two light longitudinal stripes. Males and females may be easily distinguished by the more powerful build of the male. Along each side of the flanks there is a row of large blue spots. The rear half of the underside of the body is ochre to orange or even flesh coloured.

On the grey to grey-brown ground colour the females and juveniles always have two light longitudinal stripes; there are usually also very insignificant light stripes on the flanks. On the light throat, females and juveniles usually have a dark backwards-pointing v-shaped mark.

Habitat: All imaginable areas that provide refuges and have ground vegetation. Dry stone walls, gardens, and plantations.

Lifestyle, husbandry, and reproduction: These lively lizards are best kept in pairs in a dry vivarium with a ground area of at least 80 × 50 cm. A layer of sand some 5 to 6 cm deep should be provided as substrate. By preference, these lizards eat overripe banana, tomato, and the fruits of prickly pear cactus *(Opuntia sp.)*. They should also occasionally be given house crickets, field crickets, and other insects. A shallow bowl of fresh drinking water should always be available.

It is advisable to keep animals of equal size together and to interchange them occasionally until a harmonious pair is found. A hibernation period of several weeks at 10 °C is advantageous for breeding. During the breeding season the female may lay two clutches, each of five to six eggs. This small number of eggs is balanced by the fact that the eggs and hatchlings are of a considerable sizes. On average the eggs of *Gallotia g. galloti* measure 18.8 × 12.1 mm. At temperatures between 28 and 30 °C the young hatch after 83 to 90 days. The young hatched by Bischoff (1985) had an average SVL of 33.1 mm and a tail measuring 70.3 mm. The young should be fed on small crickets and other insects and are not difficult to rear.

Gallotia galloti caesaris (Lehrs, 1914)
Hierro Canary Island Lizard

Distribution: Hierro.
Description: With a SVL of 8 cm, males usually reach a total length of 25 cm. Females remain somewhat smaller, at 20 cm tail length and 6.6 cm SVL. According to Bischoff (1985), some specimens may grow even larger. The entire upper surface of adult males is a uniform black, sometimes showing a hint of a lighter longitudinal stripe. Transverse bands never occur and there are no large blue spots on the cheeks or flanks. Occasionally, however, some small blue or greenish flecks may be found on the flanks. Frequently the side scales on the underside of the body are largely blue in colour. Females and juveniles may be immediately recognised by their significant light longitudinal stripe. The throat is black.
Habitat: "On the island of Hierro it is difficult to understand how these lizards survive on lava fields so completely devoid of vegetation" (Rogner, 1982).
Lifestyle, husbandry, and reproduction: See *Gallotia g. galloti*.

Gallotia galloti eisentrauti Bischoff, 1982
Eisentraut's Canary Island Lizard, Northern Teneriffe Canary Island Lizard.

Distribution: Northern Teneriffe.
Description: Males reach a total length of up to 43 cm of which a good 14 cm is SVL. Females remain smaller at 34 cm (SVL: 11 cm). It is only the males from Roque de Garachico area on the north coast that reach the greatest length.

Gallotia galloti eisentrauti is the most colourful of all subspecies of the Canary Island Lizard. The head, throat, neck, breast, and front of the flanks, including the forelegs, are a deep black. The colour changes towards the rear becoming grey to reddish brown. Numerous yellow to greenish, irregularly spaced transverse bands, which are edged in black, start at the back of the head and run over the entire body to the joints of the hindlegs. There is a row of medium-sized, rarely large, spots along each flank. Along the side scales of the belly there is a row of irregularly spaced blue flecks. The cheeks are largely blue to violet in colour. The rear of the underside is mainly orange to flesh coloured.

The grey to reddish brown females have a dark brown stripe and a row of small blue flecks along each side. On the back there are numerous irregularly spaced, yellow transverse bands. Longitudinal stripes may occur—if at all—in the neck area. Juveniles are very similar to females, only somewhat more red and a little lighter. Females and juveniles with longitudinal stripes are extremely rare. The characteristic V-mark may be found on the throat.
Habitat: Rock piles, dry stone walls and similar places that provide refuges. Frequently found at the edges of gardens and plantations. Because northern Teneriffe

Gallotia galloti palmae

has more vegetation than the south, the animals live together harmoniously in colonies.
Lifestyle, husbandry, and reproduction: See *Gallotia g. galloti*. The eggs measure on average 17.1 × 10.7 mm. An egg measuring 18 × 11.5 mm had increased to 24 × 17 mm after only 2 months (Bischoff, 1985d). Hatchlings have a SVL of around 35.3 mm and a tail length of 82.6 mm.

Gallotia galloti gomerae (Boettger & Müller, 1914)
Gomera Canary Island Lizard

Distribution: La Gomera.
Description: Males have an average total length of 28 cm and a SVL of 9.3 cm. In the smaller females these measurements are 24 and 8 cm respectively. Colouring and pattern are almost identical to those of *Gallotia galloti caesaris*, only the side scales on the underside have smaller blue flecks. In this species the females and juveniles also have a black throat.
Habitat: See *Gallotia g. galloti*.
Lifestyle, husbandry, and reproduction: See *Gallotia g. galloti*.

Gallotia galloti palmae (Boettger & Müller, 1914)
Palma Canary Island Lizard

Distribution: Las Palmas.
Description: Males have an average total length of 32 cm and a SVL of 10.5 cm, or at most 11.4 cm. At around 27 cm (SVL: 9 cm), the females are somewhat smaller and less heavily built than the males. The head and sides of the neck of older males are a deep black colour. Along the medium brown back there are numerous irregular transverse bands of a yellowish or greenish colour. These bands are more obvious on the latter one-third of the body. Along each flank there is a row of medium-sized blue flecks. The cheeks and frequently also the throat are a vivid blue to violet. Because of the colouring and pattern on their back, females and juveniles are reminiscent of *Gallotia g. caesaris. Gallotia g. gomerae* or *Gallotia g. eisentrauti*. However, since they do not have the black throat of these species—having instead the V-shaped marking—they are quite easily distinguished.
Habitat: See *Gallotia g. galloti*.
Lifestyle, husbandry, and reproduction: See *Gallotia g. galloti*.

Gallotia stehlini (Schenkel, 1901)
Giant Canary Island Lizard

Distribution: Gran Canaria.
Description: This lizard is the largest species of the genus *Gallotia*, and indeed perhaps the largest of all present-day Lacertids. Males may reach a total length of up to around 80 cm, but are usually somewhat smaller. Females have an average SVL of around 20 cm. The head is well set-off from the body. In both sexes the upper side of the body is more-or-less dark red to grey-brown in colour. The foremost section of the body is very dark, sometimes almost black. On many specimens there is a hint of somewhat lighter, irregularly-spaced transverse bands along the back. The lower reaches of the sides of the head, the cheeks, and parts of the sides of the neck are vivid yellow to orange. The underside of the body of these lizards is yellow to orange in colour, frequently darkly clouded. There is a backwards-pointing V-shaped mark on the throat. Along the back, juveniles have a white unbroken longitudinal stripe on an olive-brown background. There is, in addition, insignificant dark-light transverse banding. The underside is white flecked.
Habitat: Over the entire island, less frequently in woodland, but most common on banana plantations, in tomato fields, and on refuse dumps.
Lifestyle, husbandry, and reproduction: Throughout their life in captivity these lizards remain extremely shy and timid. I keep them under the same conditions as *Gallotia g. galloti*. In the wild their flight distance is 10 to 15 m, but juveniles will allow a person to approach much nearer before fleeing into a refuge. This species requires an extremely spacious vivarium with numerous hiding places and refuges. The males start the courtship procedure by compressing the sides of the neck and

pointing the snout downwards. This is followed by frantic head bobbing before lunging at the female and seizing her by a fold of skin on the neck and then completing the copulation. In Bischoff's case (1974) and my own (1979/80), the females laid clutches of up to eleven eggs measuring around 26 × 14.6 mm. During incubation the eggs increased considerably in size and the young hatched after 65 to 72 days at an average temperature of 28 °C. The females laid second clutches within a year. The hatchlings had an average length of 94.3 mm, of which 42.2 mm was tail. The willingness to mate is enhanced if the breeding season is preceded by a winter rest period of several weeks at 10 °C. According to the size of the lizards, the rest period may last from 6 to 10 weeks. The young are extremely easy to rear when fed on a varied diet of insects.

Ichnotropis Peters, 1854
Rough-Scaled Lizards

The representatives of this small genus do not have a collar. They have especially rough scales on the head and large spiny overlapping scales along the back. Below the toes there are heavily keeled lamellae. The tail is relatively long.

All *Ichnotropis* species are ground dwellers capable of moving very quickly over the sandy ground of the savannahs. Some species only live for 1 to 1 1/2 years. These species are however, sexually mature at 5 to 8 months old and die soon after laying one (rarely two) clutch(es) of eggs.

Ichnotropis capensis Smith, 1838
Rough-Scaled Cape Lizard

Distribution: Northwards to Zululand and Transvaal, southern Mozambique through Botswana, Zimbabwe and northeastwards to Namibia, Angola, and Zambia.
Description: These medium-sized lizards may reach a total length of up to 20 cm, but are usually fully grown at around 16 to 18 cm. At the centre of the body there are 28 to 42 rows of scales. The prenasal scale is undivided; in front of the infraoculars there are four labial scales. The dorsal scales are heavily keeled and overlap. Each of the hindlegs has 9 to 14 femoral pores. Along each side of the pale grey-brown back, juveniles have a white stripe. Adults have a uniform grey to yellowish brown back, along each side of which runs a narrow dorsolateral white stripe which may be bordered by black flecks. Along each flank there is a broad black stripe bordered below by a white stripe. The underside is white. When males are in breeding condition the white stripes, as well as the chin and throat, turn bright yellow. A vivid red stripe also appears on the lower flanks.
Habitat: Dry to relatively moist savannahs.
Lifestyle, husbandry, and reproduction: During the day these lizards hunt for termites and other insects. The breeding season is between October and December. Eggs are laid soon afterwards. Females dig a tunnel some 10 to 20 cm long in the

loose ground and it is here that they lay between three and nine eggs. The young hatch after 56 to 77 days measuring between 50 and 70 mm. The young grow very quickly and with a length of 15 to 16 cm after only 6 to 8 months are sexually mature. In captivity they should be kept under the same conditions as *Eremias* but without a hibernation period.

Lacerta Linnaeus, 1758
Lacertids

With over 30 species, the Lacertids may be found in Europe, Western Asia, and Africa. They live in very divergent biotopes and thus make divergent demands regarding temperature and humidity. They are at home on bushland, the edges of dry woodland, stony to rocky regions, and steppes.

Amongst Lacertids the collar is always easily distinguishable and consists of enlarged scales. The dorsal scales are small and are either not, or only very slightly, overlapped. The ventral scales are large, rectangular, and arranged in more-or-less transverse and longitudinal rows. Within this genus there are two very distinct groups: *Lacerta agilis, Lacerta viridis, Lacerta trilineata, Lacerta schreiberi*, and *Lacerta striata* belong to the group of Emerald Lizards as do *Lacerta lepida* and *Lacerta princeps* which differ slightly from the previously named species.

All other species belong to the group of Forest Lizards, even the so-called "Rock Lizards" because they do not differ greatly from them. The lizard *Lacerta australis* which is at home in South Africa was afforded its own genus some time ago (Arnold, 1989).

Lacerta agilis Linnaeus, 1758
European Sand Lizard

Distribution: In Europe from southern England, France through Denmark, southern Sweden to Lake Baikal in the east. Southwards they reach to the Pyrenees and the northern edge of the Alps, through the northern part of the Balkan Peninsula and the Caucasus to Central Asia.

Description: This medium-sized lizard, which in comparison to other species has a very short tail, has a SVL of around 11 cm. The tail is around 1 1/2 times as long. Along the centre of the back there is a row of narrow scales that are considerably higher than the outer, wider dorsal scales. Often the small nasal scale does not touch the nostril. The grey to brown upper side has one to three complete or broken longitudinal stripes between which there are rows of black spots. The flanks also have black and white flecks. Males have vivid green flanks and throat, especially during the breeding season when even the entire dorsal surface may be green. The uniform brown juveniles have small white spots along their sides. These frequently form three light longitudinal stripes.

Habitat: Within their enormous distribution range these lizards may be found on

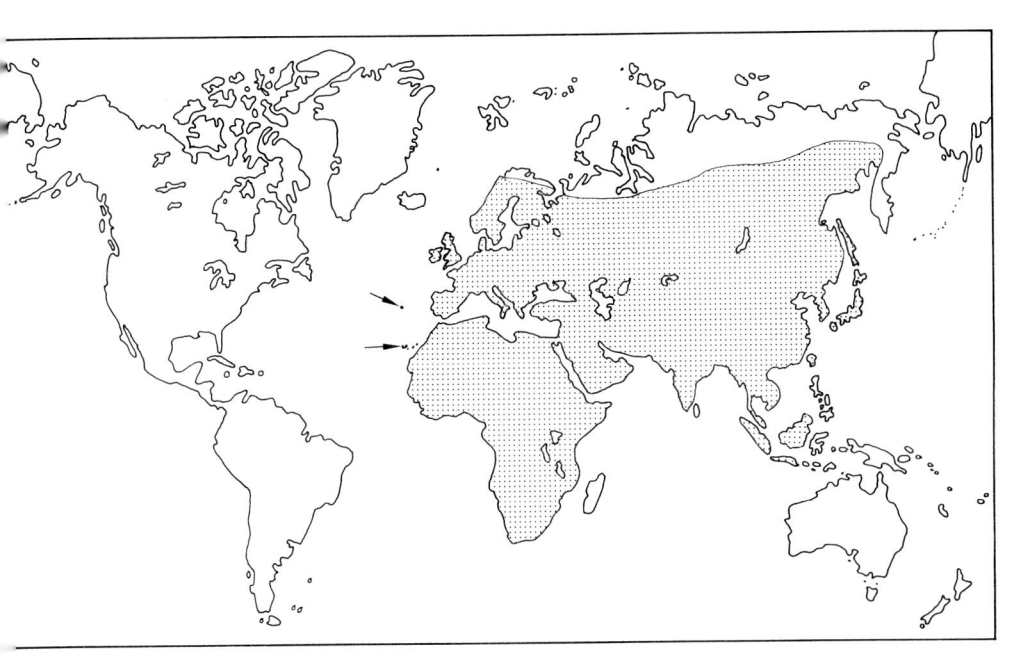

Distribution of the Lacertidae

heathland, forestry plantations, sunny hillsides, roadsides, railway embankments, forest edges, amongst bracken, in moist meadows, and even on the banks of streams with sparse vegetation or flooded areas.

Lifestyle, husbandry, and reproduction: In general, all *Lacerta* species require more moist conditions than their close relatives *Podarcis*. For this reason they are frequently found near water. With a south-facing slope in the garden, these lizards may spend the summer in an outdoor enclosure or outdoor vivarium. They require a sandy, gravelly substrate with rocks or tree roots to provide adequate and sufficient hiding places and refuges. A shallow bowl of fresh water should always be available. The outdoor enclosure or vivarium should be covered with glass or a transparent plastic sheet. Because of the natural climatic variations and unfiltered sunlight to which they are accustomed, these lizards are exceptionally easy to keep and breed in an outdoor enclosure. In a vivarium however, where light and temperature are often held constant, they will often succumb. To be able to keep a check on them, even those lizards that are normally housed in an outdoor enclosure should be overwintered in a cool cellar at 4 to 6 °C. During the transition period from winter to summer, the lizards may be kept in an indoor vivarium (see *Lacerta brevicaudata*).

In the wild, Sand Lizards feed on various insects and their larvae. They will also eat earthworms and small snails. In Central Europe these lizards hibernate from September or October until March or April.

Lacerta agilis

During the breeding season, males will fight one another, usually resulting in a male living with one female in a territory that he will defend vigourously. According to the geographic origin the females lay once or twice each year, producing from 4 to 15 eggs measuring around 7 to 11 × 11.5 to 17 mm. The young hatch after around 7 to 10 weeks. The length of incubation is determined by the temperature. Jensen (1982) experimented with the eggs of *Lacerta agilis* and found that at temperatures between 22 and 31 °C the young took between 30 and 60 days to hatch. Sand Lizards are sexually mature after their second hibernation.

Lacerta agilis brevicaudata Peters, 1958
Short-Tailed Sand Lizard

Distribution: In Central Transcaucasia, in the Armenian mountains and neighbouring areas of northeastern Turkey.
Description: These Sand Lizards appear to be rather compact and have only a very short tail. Adult females are brown with irregular black spots, whilst adult males are

Lacerta agilis brevicaudata

green. The usual original spots are lost with increasing age. In addition, uniform brown females and green males are also known.

The underside of females is white with small black spots whilst that of males is green. The Short-Tailed Sand Lizard may reach a total length of around 28 cm.

Habitat: The south-facing slopes of the Caucasus, bush steppe regions between 800 and 2,000 m above sea level.

Lifestyle, husbandry, and reproduction: See *Lacerta agilis*. These lizards are very peaceful and should therefore not be housed with other species. Liesack (1988) kept this species in pairs in outdoor enclosures measuring 60 × 60 × 30 cm. These enclosures were covered with fine gauze. During spells of bad weather the lizards were transferred to indoor vivaria in which they also spent the period from September to December or early January. Finally, they were given a 6- to 8-week winter rest period before again being placed in the outdoor enclosures. It was here that they mated and eggs were laid. The 60 × 80 cm vivaria were simply furnished. One corner contained light garden earth planted with *Saxifrage* whilst the remainder of the substrate was sand. A few flat rocks and a drinking bowl completed the furnishings. These lizards should be kept under relatively moist condition (Liesack, 1988). They require daytime temperatures between 25 and 30 °C, localised beneath a spotlight to around 25 °C. These lizards are sexually mature at 2 years of age. The females invariably laid their eggs in the moist sand near the drinking bowl. Eggs should be transfered to an incubator and should be artificially incubated at high atmospheric humidity. At temperatures between 25 and 30 °C the young take between 36 and 43 days to hatch. Even the young may be placed in outdoor enclosures during favourable weather, although care should be taken that they do not become overheated. They will not hesitate to eat all manner of insects, but care must also be taken that they do not grow too quickly or become obese.

Lacerta "armeniaca" Mehely, 1909
Armenian Sand Lizard

Distribution: Northern Armenia, northwestern Azerbaijan, southern Georgia, and northeastern Turkey (1,700 to 2,200 m above sea level).
Description: These medium-sized lizards reach a SVL of over 7 cm and have a brown to olive-coloured upper side. The dark flanks have irregular scrolls and spots that form a reticulated pattern. In spring, males are particularly conspicuous because of their yellowish green to vivid green dorsal patterning. Below the flanks there are rows of deep blue scales that delineate the ventral side.
Habitat: Gravel pits, rocky hillsides, and roadside verges.
Lifestyle, husbandry, and reproduction: See *Podarcis sicula*. Amongst the Caucasian Sand Lizards there are some parthenogenic forms, the populations of which consist of only females that reproduce without mating. One of these forms is *Lacerta "armeniaca."* The eggs develop without having been fertilised and produce only females. Males occur only very rarely and are invariably infertile. They usually degenerate during embryonal development. If a female of the parthenogenic form should mate with a male from a normal bisexual population, this results in sterile triploid offspring. Because the parthenogenic forms are sometimes regarded as individual species, their special status is indicated by the quotation marks.

In Wittig's case (1987), a female *Lacerta "armeniaca"* was kept for 3 years in various vivaria. It was only after being placed in an outdoor vivarium in spring that the body assumed a green sheen after which eggs were laid. Unfortunately the eggs failed to develop. Successful breeding was however achieved in that October. Wittig removed two young from the outdoor enclosure. Unbeknownst to him the eggs had been laid and hatched there. The hatchlings developed well but grew much more slowly than the young of *Lacerta saxicola* and *Lacerta raddei* which Wittig was rearing at the same time.

Lacerta bedriagae Camerano, 1885
Tyrrhenian Sand Lizard

Distribution: Corsica, Sardinia, and neighbouring islands.
Description: A strongly compressed lizard with a pointed head that reaches a SVL

of up to 8 cm. The tail is 1 1/2 to 2 times as long. The fairly wide cheeks make the head appear even more pointed. The yellowish green, greenish, brownish, or blackish grey ground colour is overlaid on the back and flanks by a blackish or brownish reticulated pattern. Nevertheless, at a distance, these lizards appear to be a uniform green or brownish black without any markings. The underside has a greyish white, yellow, green, or reddish tone broken partially by darker markings.

Habitat: These Sand Lizards live in the mountains at altitudes of 500 to 700 m above sea level. In northern Sardinia they also live on the seashore. In Corsica they live at altitudes of up to 2,700 m and in the remainder of Sardinia up to altitudes of 1,800 m. The actual habitats are rocky, stony, and somewhat moist areas with sparse vegetation.

Lifestyle, husbandry, and reproduction: This species is well adapted to lower temperatures and may even be seen near the snow line. At higher altitudes they undergo an extremely long hibernation period of 5 to 6 months. The breeding season begins soon afterwards. A normal clutch consists of three to six eggs each measuring around 6 × 12 mm. According to geographic origin and altitude, the young hatch during the period of July to September.

Because of their adaptation to lower temperatures these lizards may be safely kept and bred in an outdoor enclosure. When given a wide variety of insects, this is an extremely easy species to keep. They do however require large piles of rocks. When the temperature becomes too high the lizards will retreat into their refuges. In the case of Forman & Forman (1981), when kept in an outdoor enclosure these lizards were only seen during the early hours of the morning and after 1800 h in the evening, especially on bright sunny days.

Lacerta cappadocica Werner, 1902
Cappadocian Lizard

Distribution: Turkey and neighbouring regions. The nominate form lives in the Kilikian Taurus; *L. c. muhtari* east of the Euphrates and in northeastern Iraq; *L. c. schmidtlerorum* in the areas around Diyarbakir and Viransehir; *L. c. urmiana* east of Siirt-Cizre as well as in northeastern Iraq and northwestern Iran; *L. c. wolteri* lives in the area around Amanus.

Description: This lizard, which has an extremely pointed head, may reach a total length of around 23 cm. Their ground colour is light beige to brown. There are two light stripes along the back. These stripes run together on the tail which may sometimes be pale blue in colour. There are dark stripes and flecks along the centre of the back whilst the flanks have a unique pattern consisting of large blue spots bordered in black. Females are less attractively coloured than males.

Habitat: As a cliff dweller on the northern Anatolian to northwest Persian mountains, these lizards live on sunny cliffs as well as on buildings and bridges. In contrast to their relatives, *L. danfordi*, which inhabit the same area, these lizards live in warmer and drier habitats.

Lacerta bedriagae (above), Lacerta cappadocica (below)

Lifestyle, husbandry, and reproduction: According to Franzen (1990), these timid and very swift lizards live only in very small communities. They do not live in large colonies as is the case with *L. rudis*, *L. parvula*, and *L. clarkorum*. In Cappadocia, I too did not find any large colonies.

These lizards prefer to mate in the numerous rock crevices in their habitat. Franzen observed the mating of *L. muhtari* at the beginning of April at Birecik on the Euphrates.

These lizards require tall, spacious vivaria that enable them to climb at will. Several stacked rocks will provide hiding places. The upper reaches of the stacked rocks should have additional heating provided by a spotlight. During the day the temperatures should be very high, but a dramatic drop in temperature is vital at night. The vivarium should be sprayed daily. In all other respects these lizards should be kept as other cliff-dwelling species.

Lacerta caucasica Mehely, 1909
Caucasian Lizard

Distribution: Over the entire Caucasus region to northeast Azerbaijan and in the eastern Caucasus foreland (Dagestan).
Description: With a total length of only 184 mm this lizard is the smallest European species of the *L. saxicola* group. Apart from its diminutive size, it is differentiated from the partially sympatric *L. rudis* by its scalation.

On the upper side the colour varies from greenish or yellow-green through olive tones to brownish grey or a grey colour. A dark longitudinal stripe extends down the centre of the back, and from the neck downwards encloses a further series of irregular black spots or flecks. From the rear of the eye and through the temporal region there are parallel dark longitudinal stripes. The dark dorsal stripe is frequently absent or consists only of a longitudinal row of irregular spots. On the underside, these lizards are yellow, greenish yellow, or white.
Habitat: This species inhabits rocky hillsides, piles of rocks on the banks of mountain streams, stone walls, or gorges in the central and upper reaches of the mountains into the subalpine or even alpine zones. They are also found on rocks in mountain pastures or between tree trunks in forest clearings. Hollows beneath rocks, rodent burrows, and rock crevices provide refuges. On the north-facing slopes of the Great Caucasus they live at altitudes between 1,200 and 2,200 m; on south-facing slopes between 1,000 and 3,200 m. In Dagestan the subspecies *L. c. dagestanica* may be found at lower altitudes.
Lifestyle, husbandry, and reproduction: In the wild these lizards feed mainly on hymenoptera and other arthropods. Less frequently they will also eat earthworms and snails. On south-facing slopes of the Great Caucasus, mating takes place at altitudes between 1,500 and 2,100 m from May until early June. In Dagestan mating takes place earlier. Around 2 months later the females lay between two and four

eggs measuring around 13.5 to 14.5 × 6.5 to 8.5 mm. The young hatch after 6 to 8 weeks measuring between 22 and 27 mm.

Caucasian Lizards may be easily kept in an outdoor enclosure that contains several piles of rocks. On extremely hot days however, the lizards will retreat into their refuges and will only appear again in the early hours of the morning or late hours of the afternoon. Occasionally they will even remain in their refuges for several days. In late autumn the lizards should be placed in a cellar at 4 to 5 °C for a 2 to 3 month hibernation after which they are again placed in an outdoor enclosure. These lizards will thrive at temperatures between 20 and 25 °C, cooler at night. Beneath a spotlight there should be a localised temperature of around 35 °C. Ultraviolet light will often bring these lizards into breeding condition. Eggs are normally laid in a slightly moist place in the vivarium, i.e., beneath the drinking bowl. At the end of May the lizards may again be placed in the outdoor enclosure.

Lacerta clarkorum Darevsky & Medvedyeva, 1977
Clark's Caucasian Lizard

Distribution: The eastern, Turkish coast of the Black Sea and neighbouring Georgia.
Description: From a total length of up to 19 cm, two-thirds of this is tail. Males may be easily recognised by their bright yellowish green, brown to green dorsal colouring. Females are usually grey-brown in colour. The flanks of both sexes are brown. Along the back there are irregular rows of spots that may run together to form large patches. These are the same colour as the flanks. The underside is pale yellow to bright yellow. This species is distinguished from the similar *Lacerta rudis* by the dorsal scales that are always smooth and shiny.
Habitat: In the moist forest regions of the Black Sea, on roadside verges, woodland paths, cliff faces, piles of wood, and stone wells. This species prefers moist places and well-established structures.
Lifestyle, husbandry, and reproduction: These lizards scurry around their territory for only a few hours each day searching for food. Franzen (1991) observed them eating earthworms, spiders, centipedes, and a significant number of elderberries. It is highly possible that this species is active throughout the year, especially at lower altitudes and under favourable climatic conditions.

Clark's Caucasian Lizard requires the typical vivarium for cliff-dwelling lizards. They must be able to climb freely on stacked rocks or branches. The vivarium should be kept slightly moist by spraying daily. It would appear that these creatures have a great need of moisture and liquid. A spotlight should be in operation for only a few hours each day, and the temperature should never be above 30 °C. At night the temperature should be reduced to 15 to 20 °C if the lizards are to remain healthy. Even if in the wild this lizard appears to be active throughout the year, in the vivarium they should spend the winter months at reduced temperatures with the spotlight being switched on only very occasionally. According to Franzen, total hibernation is not necessary.

Lacerta danfordi Günther, 1876
Danford's Lizard

Distribution: Aegean Islands (Symi, Icaria, Samos, Rhodos, and Pentanisos) and in the southwest of Turkey.
Description: Danford's Lizard reaches a total length of only a little over 20 cm. Their colouring and patterns vary considerably. A light central stripe extends from the back of the head, along the back to the tail. On each side of this central line there is a dark flank stripe. The entire dorsal surface has an irregular marbled pattern. In juveniles the blue-green tail is particularly noticeable.
Habitat: Stony areas, often on stone walls, even in dry areas.
Lifestyle, husbandry, and reproduction: As for *Podarcis muralis* and other wall lizards.

Lacerta derjugini Nikolsky, 1898
Derjugini's Lizard

Distribution: Western and central Caucasus in the area of Krasnodar and in northeast Turkey.
Description: Derjugini's Lizard reaches a total length of 14 to 15 cm of which the tail is 1 1/2 times the SVL. The tail has alternating narrow and broad segments. On the underside of the thighs there are only 9 to 13 femoral pores. The light brown to olive or grey-brown back is covered with irregularly arranged dark flecks. The upper side of the tail is lighter than the back. A large dark stripe, interrupted by light flecks extends along each flank and down the sides of the tail. The underside is a yellow to greenish colour. In males the side row of ventral scales has blue spots. There are often one or two blue spots on the shoulders. The young are easily distinguished by their vivid green tail.
Habitat: This species prefers a somewhat moist biotope especially in forest regions and the mountains from 1,800 to 1,900 m above sea level. These lizards frequently live amongst the undergrowth at the edges of forests and on the banks of rivers. They have even been found on the seawalls of the Black Sea.
Lifestyle, husbandry, and production: Derjugini's Lizard feeds mainly on insects and spiders. They will however also eat woodlice, snails, and earthworms. Under favourable climatic conditions, they will leave their hibernation quarters as early as the end of February or March. It is however more usual for them to appear at the end of March or early April. They mate in May. From June to July the females lay two clutches of eggs each containing between four and eight eggs measuring 9 to 11 × 7 to 7 mm. It is not unusual for several females to lay their eggs in the same place,

Lacerta clarkorum (above), Lacerta danfordi (below)

Lacerta derjugini (above), Lacerta graeca (below)

such as amongst the roots of fallen trees, in rock fissures, or in clefts in tree stumps. The young hatch during the period from late June until early August and have a SVL of 19 to 26 mm. They are sexually mature after the second hibernation. In appearance and lifestyle these lizards are very reminiscent of the European Viviparous Lizard, *Lacerta vivipara*.

Lacerta graeca Bedriaga, 1886
Green Rock Lizard

Distribution: Limited to the Pelopponese Islands of southern Greece, usually from altitudes of 400 m upwards.
Description: With a SVL of around 8 cm and a tail that is twice as long, this species is one of the medium-sized lizards. The ground colour of the flat rump is a vivid grey-brown to bronze-brown with irregular black spots on the upper side. Along the flanks there are numerous light spots. The shoulder area has one, at most two, blue spots. The ventral-edge scales of males frequently have blue spots. The orange or yellow underside usually has black spots.
Habitat: Rock and cliff lizards that prefer to live in the vicinity of water, especially in moist places near streams.
Lifestyle, husbandry, and reproduction: In the wild this species hunts mainly on cliffs where it catches insects and other arthropods. Very occasionally they will also hunt on the ground. When climbing these lizards move very slowly, carefully dragging the tail behind them.

Gravid females, but not any juveniles, are found in the wild during June. Obviously the young do not hatch before August. In the case of Langerwerf, on May 12 a female laid five eggs in the vivarium. From these eggs, which were incubated at 28 to 29 °C, the young hatched between June 19 and 22. Such a short incubation period could not however be expected under natural climatic conditions. After laying eggs, Langerwerf's female was very aggressive towards other lizards of the same species, as a result of which the entire breeding group died. None of the wounds inflicted by the female could be healed and this led to death from necrosis after only a few days. These lizards become sexually mature after their first hibernation. They should be kept under the same conditions as other Rock Lizards. During the summer months it is advisable to keep this species in an outdoor enclosure.

Lacerta horvathi Méhely, 1904
Croatian Lizard

Distribution: Western Slovenia, western Croatia. Also in the Julian Alps, partially on Italian territory.
Description: The Croatian Lizard may reach a total length of 16 to 18 cm, of which 6.5 cm is the SVL. Like the body, the head is fairly compressed. On the light grey to

brown-grey back there is often slightly darker speckling or a central stripe delineated by dark brown stripes on the flanks. The edges of these stripes are wavy. They begin on the sides of the head and run down to the tail. The underside is a yellowish to greenish colour.
Habitat: Mountains, usually at altitudes from 500 to 2,000 m above sea level. Preferably on moist hillsides with stony, overgrown ground and subject to intense sunshine.
Lifestyle, husbandry, and reproduction: In the wild these lizards frequently live together with *Podarcis muralis* and *Lacerta vivipara*. This is not a very swift species, but it is extremely timid. In the vivarium they may be fed on the usual insects. In Méhely's case, a female in a vivarium laid four eggs on July 7. The eggs measured 13.6 to 14 × 7.5 mm and were an elongated elliptical shape. In his vivarium, Bischoff found three clutches, each of three eggs on June 14. At an incubation temperature of 27 °C the young hatched after 37 to 41 days measuring around 6.2 to 7 cm in length. A second clutch is possible. These lizards may be kept in the same way as others that are not sensitive to cold and can be placed in an outdoor enclosure at the beginning of May when cold spells are no longer likely.

Lacerta jayakari Boulenger, 1887
Oman Lizard

Distribution: Oman Mountains along the west and southwest coast of the Gulf of Oman, from the Musandam Peninsula in the north to Hajar ash Shargi in the southeast (Bischoff, 1981).
Description: Both sexes may reach a SVL of over 16 cm. The tail is around 2.4 to 2.8 times the SVL. The long powerful legs of this slender lizard have toes that are distinctly bent on the hindfeet and appear to be bent upwards on the forefeet. There are significant skin folds on each side of the neck. The beige-to-light-brown upper side has irregular reddish brown spots that usually run together to form transverse bands. In addition there are very small dark-brown-to-black spots that, especially in older males, form a scroll-like pattern. Older males also have turquoise blue spots on the head and body.
  Males may be easily recognised by the broader, more powerful head, a thicker tail root, and larger femoral pores. They are also heavier than the equally large females (Bischoff, 1987).
Habitat: Sufficiently moist wadis (dried-up river beds) on which palms and citrus trees grow. Oman Lizards are also found on cliff faces and in excavation areas. In the wadis they live on the ground amongst dense *Oleander* growths on the banks of irrigation canals and on stone walls.
Lifestyle, husbandry, and reproduction: These lizards seek refuge between rocks and are expert climbers. At temperatures between 33.5 and 40 °C, they are active

and may be seen throughout the day. In the vivarium Oman Lizards are very lively in the morning. Around noon they will often retreat to cooler places, appearing again around 1600 h. In principle they should be kept in the same way as the Ocellated Lizard *(Lacerta lepida)*, but they do not require a hibernation period. However, a rest period during which the temperature is reduced by around 3 to 5 °C will do no harm. Being omnivorous, the Oman Lizard in the wild eats not only all manner of insects, but also vegetarian food such as leaves, fruit, and seeds. In the vivarium they will readily eat mealworms, crickets, flies, nestling mice, raw meat, fruit, and yoghurt.

It is likely that in the wild this species breeds throughout the year. Bischoff (1981) saw his Oman Lizards mating numerous times in the vivarium. Like *Gallotia* species, the male bites the female's neck before the actual copulation takes place. Three to 4 weeks later the female lays between seven and eight (rarely nine or ten) eggs. Even when laid the eggs are large, measuring on average 24.1 × 15 mm. During incubation they increase considerably in both size and weight. After 12 weeks they can have average dimensions of 29.3 × 24.2 mm and weigh around 9.3 g. At temperatures of around 27 to 30 °C and 100% relative atmospheric humidity, the young hatch after 116 to 119 days measuring around 17 to 18 cm in length. At a constant temperature of 30 °C the eggs hatch after only 91 to 94 days. At even higher temperatures, the eggs hatch after 75 to 85 days (Langerwerf, 1984). During the first weeks of life the juveniles are extremely susceptible to stress and must not be handled. They are also very voracious and therefore easy to feed.

Lacerta laevis Gray, 1838
Syrian Lizard

Distribution: Southeast Turkey (Central Taurus and Hatay), northwest Syria, Lebanon, Israel, and western Jordan (in the south to Petra).
Description: Maximum SVL of 80 mm with the tail being 2 to 2.3 times as long. On the uniform brown or grey back there are sometimes dark flecks. Along the sides there is a wide dark stripe often with white speckling. The underside is white, intense red, or greenish blue to intense blue. There are three recognised subspecies.
Habitat: The distribution range of this species is limited to the Mediterranean region. As with other representatives of the *Lacerta danfordi* complex, they show a distinct preference for more moist habitats. The preferred temperature of *Lacerta laevis* from Palestine is around 31.7 °C.
Lifestyle, husbandry, and reproduction: These lizards climb on cliffs, trees, roadside slopes, and the fringes of woodland paths. Large colonies are found in areas of dense vegetation and high humidity.

The Syrian Lizard does not prefer surfaces as steep as those favoured by other Rock Lizards and may be frequently found on the ground. At lower altitudes these lizards are active throughout the year, but at higher altitudes in the mountains they hibernate for up to 5 months. In lowland areas mating takes place from December to June, whilst in the mountains the breeding season is from March to July. In Osenegg's case (1989), a wild-caught gravid female *Lacerta laevis troodica* laid eggs on April 20, 1987. Further matings took place during May and June. The first clutch contained four eggs. At 29 °C and 100% humidity, two young hatched after 47 days: one male and one female with a SVL of 25 mm and a tail some 55 mm long. At the age of only 6 months the new female laid her first clutch of eggs measuring 10 × 5 mm.

Lacerta lepida Daudin, 1802
Ocellated Lizard

Distribution: Over the entire Iberian Peninsula with the exception of the northern coast of Spain and the high mountains of the Pyrenees. In France they inhabit a broad band of coastal region along the Mediterranean and into Italian territory.
Description: The largest lizard of Europe has a SVL of more than 20 cm with the tail being 1 1/2 to 2 times as long. Giant specimens with a total length of 90 cm have already been authenticated. The upper side is green, or very rarely a brownish colour. Along the flanks there are usually large, deep blue spots. The underside is a yellow to green colour. Juveniles are particularly attractive having a green dorsal surface with numerous white spots bordered in black. These spots form a series of transverse bands along the body.
Habitat: Rocky and very sunny landscapes, cultivated land with sufficient hiding places. In the Pyrenees and Alps up to an altitude of 1,000 m and in southern Spain up to 2,100 m above sea level.
Lifestyle, husbandry, and reproduction: In the wild these lizards feed not only on insects but will also eat other lizards, mice, young snakes, young birds, and sweet overripe fruit. Ocellated Lizards are active from early March until late October, after which they hibernate. Males will vigorously defend a large territory against intruders of the same species.

These lizards are best kept in pairs in a vivarium that should measure at least 100 × 50 × 50 cm (Bischoff, 1985). A 5 to 6 cm layer of coarse river sand is a suitable substrate. A very stable pile of rocks will provide refuges and stout branches will allow the lizards to climb at will. During the day, this species requires temperatures between 25 and 30 °C, localised beneath a spotlight to 35 to 40 °C. At night the temperature should be reduced to around 18 to 20 °C. A rich, varied, and adequate diet is essential. Suitable foods are all manner of insects, earthworms, young mice and rats, fruit yoghurt, sweet berries, and other fruits.

Ocellated Lizards mate soon after hibernation. During the breeding season the female will often lay two clutches, each containing up to 20 eggs which should be placed in an incubator. In the wild and subjected to fluctuating temperatures, the incubation of the eggs may take up to 3 months. Because the parents will readily eat their own young, they must be reared separately.

In Rutschke's case (1989) the females laid their eggs on the dry sand and then covered them with a mound of sand. More sand was added daily so that the size of the mound increased steadily. The females also guarded their clutches for up to 6 days. As soon as they began to dig, Rutschke provided his females with a box of 20 × 15 × 10 cm containing a moist mixture of sand and peat in which the females eventually laid their eggs. At temperatures of 28 to 30 °C, the eggs hatched after 67 to 86 days producing young that on average measured 104 mm. The young should be reared in small groups. With correct husbandry and feeding, Ocellated Lizards may be sexually mature at around 1 year old. Even captive-bred animals should only be kept in pairs. Captive-bred lizards are not quite so timid as wild-caught specimens. Adults should be hibernated for 2 to 3 months; juveniles for only around 6 weeks.

Ocellated Lizards may also be kept in a large outdoor enclosure that should contain a purposely built shelter, covered with rigid, transparent plastic sheeting. The sides of the shelter should be well planted. Stacked, flat rocks, tree roots, and pieces of tree trunk provide the furnishings for the outdoor enclosure. For cooler days, an underground heating cable should be installed. Under these conditions, Ocellated Lizards have been successfully kept and bred in the lower Rhine Valley. For their own safety, hatchlings should be removed from the outdoor enclosure. An outdoor enclosure does however present one problem: in cooler summers the lizards may go into premature hibernation which can disrupt their natural cycle in the following year (Sprünken M. & Rutschke J., 1992).

Lacerta media Lantz & Cyrén, 1920
Eastern Emerald Lizard

Distribution: From Central Asia Minor in the west to the Caucasus in the east, to Israel in the south and Luristan (southwest Iran) in the southeast.
Description: This species reaches a SVL of up to 16 cm with its tail being around 1 1/2 times as long. The top of the head is a greenish colour with fine scrolling. With increasing age the upper side of the body becomes increasingly green. The throat and belly of the Eastern Emerald Lizard are greenish white to greenish yellow in colour. In males the sides of the neck are blue. Juveniles have five light stripes on a dark brown background.
Habitat: In Turkey the Eastern Emerald Lizard inhabits the continental dry-summer areas, especially in areas with much vegetation and cover such as river banks, the edges of forests, bushlands, and overgrown rubbish dumps. At Lake Van in eastern Turkey, Franzen (1990) found this species in areas almost devoid of vegetation, on alluvial land and on very limey ground with only sparse vegetation of grasses

Lacerta laevis (above), Lacerta media (below)

and thistles. In the Mediterranean region, the Eastern Emerald Lizard inhabits areas of macchia bushland.
Lifestyle, husbandry, and reproduction: See *Lacerta viridis* and *Lacerta trilineata*.

Lacerta monticola Boulenger, 1905
Iberian Mountain Lizard

Distribution: At altitudes between 1,500 and 2,000 m in northern and central Spain as well as in central Portugal. On the coast of Galicia also at lower altitudes down to sea level.
Description: From a total length of up to 23 cm, 7.5 cm is SVL. These medium-sized lizards are rather squatly built and have very variable patterning. On a green, olive, or brownish back, males usually have one or two stripes consisting of black spots. On the dark reticulated pattern along the sides there are small light spots and blue flecks. The side rows of ventral scales are often black, sometimes with blue flecks. Some specimens have a reticulated pattern over the entire body and flanks. During the breeding season, the whitish belly, which has fewer dark spots, changes to green making the animals easy to identify.
 Females are less colourful and usually have a dark stripe along a plain brown back. The stripes on the flanks have a reticulated pattern. Juveniles have a bright blue tail, but in all other respects are similar to the females.
Habitat: Inhabitants of rocky, stony areas, but also found in woods, on coastal cliffs, and stone walls.
Lifetyle, husbandry, and reproduction: Little is known about the lifestyle of this species. They are extremely sensitive to cold. Mating takes place in early April in Sierra de Cauel, later elsewhere. Before laying eggs, females dig a tunnel some 12 cm long and 2 cm wide beneath a rock. The tunnel is then usually sealed with small pebbles, sand, and pieces of vegetation. In the Pyrenees, females usually lay their 14 × 9 mm eggs during the second half of July. In captivity, at 19 to 25 °C the young hatched after 59 days. Females caught at Laguna de los Pájaros on June 25 laid between five and eight eggs between July 8 and August 23. These measured 10.7 to 14.7 × 7.2 to 9.2 mm, and developed within 35 days. Young females usually lay only two eggs whilst older females may lay up to nine. Hatchlings have a SVL of around 22 mm and a total length of 52 mm. Females are sexually mature when they reach a SVL of 60 mm. The young should be reared separately or they will be attacked and eaten by the adults (Salvador, 1984). These lizards may be kept in an outdoor enclosure during the summer months.
 A vivarium with a ground area of 50 × 50 cm is suitable for a small breeding group. The substrate should be a mixture of sand and peat that is kept constantly moist. A pile of rocks with many gaps between them should form the centrepiece of the vivarium. During the day, a temperature of up to 28 °C is required, falling at

night to below 20 °C. During the day these lizards will often seek out cooler places. They may be fed on the usual food insects.

A 4- to 8-week winter rest period will stimulate breeding, To hibernate these lizards they should be placed in a box containing slightly moist sphagnum moss and should be kept at a temperature of around 5 °C. Males usually emerge from hibernation 1 to 2 weeks before the females.

Lacerta mosorensis Kolombatovic, 1886
Mosor Lizard

Distribution: Southern Dalmatia, Herzegovina, and Montenegro.
Description: The SVL is up to 7 cm whilst the tail is twice as long. On the upper side these lizards are either dark grey, grey-brown, or olive green in colour with dark speckling or marbling. There are also specimens without markings. The back is often lighter than the flanks. The side ventral scales may be a blue tone. The unspeckled ventral surface is usually yellow or orange but sometimes also white to grey or even a greenish colour.
Habitat: In the Karst Mountains up to altitudes between 600 and 1,500 m. Prefers more moist areas such as light deciduous forests, mountainsides with boulders, juniper bushes, and springs.
Lifestyle, husbandry, and reproduction: These rare lizards leave their winter quarters very late, around the end of April or in early May, because in their natural habitat the winter may be more prolonged than at lower altitudes. At first they tend to remain in drier, warmer places, retreating to cooler and more shaded places with more vegetation later in the year. Around mid-June to early August, the females lay between four and six eggs. The young hatch in September.

In the vivarium these lizards quickly become fairly tame and will eat all the usual insects. They should be kept in the same way as other lizards that are not too sensitive to cold. Langerwerf (1983) has bred this species. In his case the lizards mated in June. No matings were seen after the end of June. From July 16 to 19, several females laid between four and eight eggs. At temperatures between 28 and 31 °C, incubation took only 23 days. This is probably the shortest incubation period of all Lacertids. On hatching the young had a SVL of 27 to 28 mm and a tail some 72 mm long.

Lacerta oertzeni Werner, 1904

Distribution: Turkey and the Greek island of Ikaria off the Turkish coast.
Description: These exceptionally wide-headed lizards reach a SVL of up to 76 mm. A characteristic of juveniles is the very prominent stripes. These consist of three dark longitudinal stripes between which run two white supercilliary stripes. The

underside is a reddish, greenish, or blueish mother-of-pearl colour. There are several subspecies.

Habitat: Often found climbing cliffs, on tree trunks, in macchia, and in light pine forests. They prefer relatively small areas with dense to medium vegetation, especially in fairly moist areas such as by the sides of streams or in the vicinity of small lakes or ponds. They may also be found amongst ruins, on dry stone walls, and on rocky mountains enclosed by forests. Various populations appear to colonise various locations, but only isolated populations are found on flat, even ground.

Lifestyle, husbandry, and reproduction: These lizards climb on rocks and cliffs but only rarely on tree trunks. At the height of summer and in winter their activities are severely limited. I was informed by Franzen that, in some populations during December and January, only a proportion of the lizards were active, even on the warmest and sunniest days.

The vivarium should be as tall as possible and should offer a great number of climbing possibilities in the form of stacked rocks that have gaps between them to provide refuges. The vivarium should be sprayed daily creating high humidity in the gaps between the rocks. A spotlight should be installed above the stacked rocks. In spring and autumn the temperature should be reduced considerably at night. To stimulate breeding these lizards should be hibernated for 1 or 2 months at around 10 °C. In all other respects they should be kept in the same way as other Rock Lizards.

Lacerta oxycephala Duméril & Bibron, 1839
Sharp-Snouted Lizard

Distribution: Yugoslavia. In southern Dalmatia, Herzegovina, Montenegro, and on neighbouring islands. From the coast inland up to an altitude of 1,500 m above sea level.

Description: The attractive Sharp-Snouted Lizard reaches a total length of only 20 cm, of which the SVL is around 6.5 cm. The relatively short body and pointed head are quite compressed. On the blue-grey, ash grey, or blackish brown body there is a light reticulated pattern and on an undamaged tail, significant alternating black and blue rings. In the mountains, very dark or even black specimens may occur. The underside is blue to blue-grey.

Habitat: Very variable areas such as rocky ridges, gravel pits, dry stone walls, and fallen trees.

Lifestyle, husbandry, and reproduction: In the wild, these very swift lizards feed on all manner of insects and other arthropods. Spiders are also especially favoured. Sharp-Snouted Lizards are much less sensitive to cold and in the wild only go into hibernation in December, with the males reappearing as early as February. During the period from March to April, these lizards live together in pairs and defend a territory vigorously against intruders of the same species. The pairs separate only

Lacerta oertzeni pelasgiana (above), Lacerta oxycephala (below)

Lacerta pamphylica

at the end of the breeding season. In June, the female lays two to four very elongated whitish yellow eggs. In the wild these hatch after 6 to 7 weeks, producing young around 5 cm in length. Their colouring is identical to that of the parents, but they have a much more brightly coloured tail. These lizards may be kept under the same conditions as those described for the Sand Lizard *(Lacerta agilis)*.

Lacerta pamphylica Schmidtler, 1975
Pamphylic Emerald Lizard

Distribution: Central south coast of Turkey.
Description: These powerfully built lizards reach a SVL of up to 12 cm with a tail up to 20 cm in length. The head is almost uniform green, often with small brown flecks. Only part of the body of males is green, usually the front parts of the back. The green colour is usually absent from females. The throat of males in breeding condition shines a vivid blue colour. The sides of the neck are ultramarine blue. The limbs and tail are brown, The underside is a whitish or yellowish colour. Juveniles have a uniform brown back (Schmidtler, 1986).

Lizards
121

Habitat: In light pine forests, on the banks of streams, amongst dense undergrowth, and on sloping hillsides.
Lifestyle, husbandry, and reproduction: In the wild, these lizards are extremely timid and will normally disappear immediately at the first sight of a human. In the vivarium they should be kept under the same conditions as other Emerald Lizards *(Lacerta viridis, Lacerta trilineata)*.

Lacerta parva Boulenger, 1887
Pygmy Lizard

Distribution: Mainly in the Anatolian Highlands of Asia Minor.
Description: In appearance this lizard is similar to the common European Sand Lizard. These animals have a SVL of up to 6 cm and a total length of around 15 cm. On a light or dark grey dorsal background there are mid-brown to blackish brown spots arranged in transverse bands as well as greyish white to pure white spots or flecks. On the flanks and shoulders males have vivid blue spots that are occasionally edged in black. The underside is white or occasionally a yellowish colour.
Habitat: Only on the Anatolian or Armenian mountain steppes at altitudes of 900 to 2,000 m, on scree slopes and rocky hillsides. Vegetation in these areas is very sparse and short.
Lifestyle, husbandry, and reproduction: Like other lizards, Pygmy Lizards will eat all manner of insects and spiders, However, the main part of their diet consists of ants. In the wild and according to climatic conditions, hibernation lasts from the end of September until the middle of April. The young are sexually mature after their first hibernation. Each year a female may lay two or three clutches each containing two to five eggs measuring around 6 to 7×12 to 13 mm. These lizards may be kept in the same way as *Lacerta agilis*, only under slightly warmer and drier conditions.

Lacerta parvula Lantz & Cyrén 1913
Dwarf Pygmy Lizard

Distribution: Eastern Turkish Black Sea coast (east of Trabzon) to Georgia (to around Bakuriani). Perhaps an isolated population in central eastern Turkey between Erzurum and Savikamis.
Description: This species reaches a total length of up to 19 cm, but is usually somewhat smaller. Of this total length, two-thirds is tail. These attractive lizards are somewhat compressed. The back is brightly coloured in various shades of brown, sometimes appearing somewhat greenish in colour. The patterning on the back often forms rows of spots or creates a reticulated pattern. Specimens without any markings are also known. The underside, especially in older animals, is often a bright brick red, whilst in other populations the underside is usually a whitish green.

Habitat: The nominate form lives on cliff faces in continental mountains which are dry in summer. Franzen described a population, the geographical and ecological situation of which was more in keeping with the coastal form *Lacerta parvula adjarica*, namely a cliff face at an altitude of 1,500 m in the vicinity of the eastern Turkish town of Savsat. There, the climate and vegetation are very similar to those in the Alps: high rainfall throughout the year, extremely cold and deep snow in winter. The landscape is mainly meadows and extensive pine forests.

Lacerta p. adjarica inhabits the subtropical areas of the Black Sea coast which have heavy rainfall. Here the lizards live at the sides of roads and pathways, on cliffs, and amongst ruins, often together with *Lacerta rudis* and *Lacerta derjungi*.

Lifestyle: It is obvious that these lizards prefer vertical surfaces. They should be kept in the same way as *L. rudis, L. clarkorus,* and other Rock Lizards.

Lacerta pater Lataste, 1880
Berber Lizard

Distribution: The Atlas lands of northwest Africa: Tunisia, Algeria, and Morocco northwards of the northern edge of the Sahara Desert and in the Tasili Mountains. The Algerian Sahara and in the Atlas Mountains themselves up to an altitude of 1,500 m above sea level.

Description: These lizards remain somewhat smaller than the related Emerald Lizard. Males reach a total length of 58 cm with a SVL of 15 to 17 cm. Females are even smaller. These powerful lizards have a less-compressed body than the Emerald Lizard. The upper side is emerald to blue-green. In eastern specimens the entire dorsal surface is green, whilst in western specimens the hindlegs and tail are brown. The underside is a greenish yellow colour and the throat a greenish white. Some males may have a blueish tinge. It is not rare for some adult males to be completely without markings, whilst others may have a small number of flecks or a reticulated pattern reminiscent of the Emerald Lizard. In females the light spotting of the juvenile pattern may still be barely visible, but they are usually without markings. Along the flanks there are sometimes three or four rows of insignificant blueish flecks. Males are very easily distinguished from females by their larger head and more powerful build.

Habitat: Frequently found in olive and fig groves, in the macchia, and in areas of sparse vegetation. They hide amongst rocks and in wood piles. The Berber Lizard is often found near water courses. In their natural habitat these lizards are subject to long hot summers and cold moist winters.

Lifestyle, husbandry, and reproduction: See *Lacerta lepida*. In the vivarium females may lay up to three, but usually two, clutches of eggs each year, each clutch containing between 9 and 14 eggs although ten to twelve is more usual. At temperatures of 27 to 28 °C the young hatch after 70 to 74 days with a total length of 114 to 118 mm.

The snout area is green. On the body there is a much smaller number of light spots than in *Lacerta lepida*, these are arranged in longitudinal rows (Bischoff, 1985).

In Langerwerf's case (1981), two males and three females lived in a glasshouse with a ground area of 30 m^2. These animals formed the basis of his breeding group. Because the first batch of young all had rickets (rachitis), the adults, especially the females, were given calcium lactate and vitamin D3 in soluble form in their drinking water during the breeding season (calcium lactate dosage: 1 teaspoon in 1 litre of water). There were no further cases of rickets. Thus in 1977, 70 young were hatched, in 1978, 186; in 1979, 215, and in 1980 around 200 young were produced. The captive-bred young lived together with other species in a glasshouse with a ground area of 50 m^2 and garden frames from 3 to 10 m^2. The first totally melanistic young hatched on October 12, 1978, together with 21 normally coloured young. These were produced from the eggs of a captive-bred female. The melanistic young proved to be a female, which during 1979 and 1980 laid several clutches of infertile eggs. It is possible that this female did not recognise normally coloured males as being of the same species. Finally, from a clutch of 17 eggs produced from normally coloured parents, further totally black young were produced.

Lacerta praticola Eversmann, 1834
Meadow Lizard

Distribution: Isolated populations in the eastern Balkans (southern Romania, northeastern Yugoslavia, east and southeast Bulgaria, the northeastern part of European Turkey) and in the Caucasus region (north and central Caucasus, Transcaucasia to northwest Armenia, Lenkoran, and the Caspian Sea). Also in northwest Iran.

Description: These lizards are very slenderly built and reach a total length of 16 cm of which 6 cm is SVL. Their colouring and markings are reminiscent of the European Viviparous Lizard, *Lacerta vivipara*. The rear edge of the collar is rather ragged and toothlike and the animals are uniformly coloured. A light brown stripe, the width of the head, runs down the body. This is often speckled with small dark spots. Down the centre of the broad stripe there is a somewhat darker, narrower stripe. There is a broad dark stripe along each flank, below which there is a narrower light stripe that begins at the eye and ends on the tail. This stripe is broken at the joints of the limbs. The underside is yellowish, but in males is often greenish and without any flecks.

Habitat: Forests, especially deciduous forests, on footpaths and roadsides, also on bushy hillsides and near marshes up to an altitude of 2,000 m.

Lifestyle, husbandry, and reproduction: This is not a particularly swift species but is an excellent climber that spends the warmest part of the day in its retreat, which is usually a tunnel that it digs itself in loose ground. These tunnels usually have two

Lacerta parvula (above), Lacerta praticola (below)

Lacerta raddei

exits. In the vivarium these lizards are normally very tolerant of both their own and other species. The substrate should be covered with fallen leaves (oak, beech etc.).

In the wild, this species hibernates from September/October until March/April. Males usually emerge from hibernation earlier than females, who only reappear several weeks later. The lizards mate from late May to late June with the female producing three to six eggs some 11 × 6 mm, usually at night during June or July. The young hatch at the end of August or in early September. Under favourable conditions in the vivarium, the female may lay a second clutch of eggs. At temperatures around 29 °C the eggs hatch after 45 to 50 days. As a precaution, the young should be reared either individually or in small groups in a large vivarium. They should be given a quite varied diet.

Lacerta raddei Boettger, 1892
Azerbaijan Lizard

Distribution: From Armenia through Azerbaijan and northeastern Turkey to northern Iran and Turkmenia.
Description: This species reaches a SVL of around 7 cm with the tail being twice as long. Males may be recognised by their thicker tail root and femoral pores. During the breeding season they are much more brightly coloured.
Habitat: Scree, cliffs, and walls with lush vegetation and strong sunlight.

Lifestyle, husbandry, and reproduction: Two pairs may be kept in a vivarium of 50 × 30 cm. The rear wall should be arranged as a cliff face. These lizards require bright lighting and regular ultraviolet radiation. During the day the temperature should be around 30 °C, falling at night to 20 °C. They feed on the usual variety of insects and other arthropods. During the summer, these lizards may be kept outdoors in their vivarium or in an outdoor enclosure on the south side of the house. A 1 or 2 month winter rest period will stimulate the breeding condition, with mating usually taking place soon afterwards. Before mating the male seizes the female firmly on the flanks. Around 3 to 4 weeks after mating, the female lays three to seven eggs in a moist place in the substrate. The 10 × 6 mm eggs take 50 to 60 days to hatch. The hatchlings are best reared individually and should be fed on a quite varied diet of insects and other arthropods. They are sexually mature after 1 to 2 years.

Lacerta rudis Bedriaga, 1886
Spiny-Tailed Lizard, Keel-Tailed Lizard

Distribution: Turkish Black Sea coast and in the mountainous hinterland of Adapazari in the west to the Russian border. Widely distributed in Transcaucasia and on the southern slopes of the Great Caucasus. This species only reaches the continent of Europe in a very isolated population in the north of the main Caucasus range.
Description: With a total length of up to 25 cm this species is the largest representative of the *Lacerta saxicola* group. Its SVL is around 8.6 cm. The body is rather thick-set. The upper side varies between green, yellow, and brown to variously intense tones of grey. The dorsal stripe consists of dark flecks on the sides along the centre of the back. This creates a light stripe on each side on which there are sometimes vivid light flecks. Three rows of black or blackish-brown spots running together form the temporals which sometimes have light ocelli. The underside is greenish, yellowish green, blueish, or whitish. Specimens without markings (the so-called "concolour" form) are occasionally found.
Habitat: In dry or slightly moist regions and in high mountains.
Lifestyle, husbandry, and reproduction: According to Franzen (written communication), these lizards are active throughout the year under favourable climatic conditions. Even on warm sunny days during December and January these lizards may be seen along the Turkish Black Sea coast in the same numbers as during the summer. They also feed during this time. It is worthy of note that at this time of year they eat mainly plant material (Franzen, 1991), such as the fruit of the elder *Sambucus nigra* which they reach by climbing. In contrast to *Lacerta clarkorus* which lives in the same area, *Lacerta rudis* also inhabits open spaces such as limestone slopes and building sites without vegetation. When kept in captivity, their origins must be determined since those specimens from high mountains require climatic conditions different from those lizards hailing from the subtropical lowlands (Franzen, 1991). Both however have a high water requirement, so the vivarium must be

Lacerta rudis

sprayed daily to achieve high humidity. The furnishings should be the same as for other Rock Lizards. Mating takes place immediately after the winter rest period, with the males emerging some 2 weeks before the females. Males are especially aggressive towards one another during the breeding season and will often undergo ritualised fights. Before mating the male always bites the female on the flank. Females lay only one clutch of two to eight eggs each year. The size of the eggs varies for each subspecies. In the largest of the subspecies they are around 16 × 8.5 mm, whilst those of the smaller subspecies are around 12.5 × 6 mm. According to subspecies the young measure 50 to 55 mm on hatching.

Lacerta saxicola Eversmann, 1834
Rock Lizard

Distribution: Southern Crimea, western Caucasus, Transcaucasia, and Asia Minor.
Description: This medium-sized lizard is somewhat reminiscent of the European Wall Lizard. It has a SVL of 7.5 cm with the tail being twice as long. On the brownish to olive-coloured dorsal surface there are numerous scrolls and flecks which may be quite insignificant or may form a reticulated pattern.

In spring, males have yellowish green to vivid green dorsal colouring, whilst on the flanks there are black-edged, deep blue shoulder spots. The side scales on the

Lacerta schreiberi

belly are also blue. Females sometimes have these blue spots but they are never so pronounced as in males. There are several subspecies.

Habitat: Dry scree slopes, cliff faces, roadsides, and similar areas. In the mountains up to an altitude of 3,000 m.

Lifestyle, husbandry, and reproduction: These excellent climbers hunt all manner of insects, larvae, and spiders. According to the altitude they may be observed in the wild from February or March until September or November. The breeding season is therefore subject to variation. This species may be kept in the same way as the European Sand Lizard *(Lacerta agilis)*.

Rock Lizards of the subspecies *L. saxicola brauneri* were kept by Wittig (1987) in an outdoor enclosure and during the spring their colour turned to a vivid green. In early November, with a ground temperature of 4 °C a clutch of four newly laid eggs was found beneath a rock. These were transferred to an incubator and hatched after 62 to 66 days. The young had a total length of 6.1 to 6.9 cm and a SVL of 2.5 to 2.7 cm. They developed well and at the age of 7 months had grown to total lengths of 13.9 to 16.1 cm.

Lacerta schreiberi Bedriaga, 1878
Iberian Emerald Lizard, Schreiber's Green Lizard

Distribution: Iberian Peninsula. Northwestern Spain (Asturia and Galicia), the central mountains in central Spain and Portugal up to an altitude of 1,800 m.

Description: These squatly built lizards reach a SVL of up to 12 cm and a total length of around 40 cm. The ventral scales form eight longitudinal rows. The quite variable colouring ranges from grey-brown through yellowish green to grass green. Along the back and flanks there are numerous small dark spots. Females are usually a brownish colour with irregular large black spots that form three stripes. There are often white eyespots along the sides. Males, and occasionally also females, have black spots on the belly. Males usually have a deep blue throat, especially during the breeding season. During the first 3 to 5 months of life juveniles are dark brown in colour, with darker spots forming three to five stripes on the flanks.

Habitat: Areas of high humidity, at the edges of streams and on scree-covered hillsides with impenetrable bramble bushes.

Lifestyle, husbandry, and reproduction: In the wild this Green Lizard eats not only all manner of insects, but also other smaller lizards. They will vigourously defend their territory. Mating takes place in May with eggs being laid in June. A clutch usually consists of 13 to 21 eggs measuring from 13.7 to 16.7 × 9.7 to 11.2 mm. When disturbed, these lizards will flee into the impenetrable bramble bushes or into rock crevices. Populations that live near water will often spring into the water when threatened. They will remain underwater for some time. Daily activity begins soon after dawn and continues until the sun no longer shines on their territory. These lizards are often found basking on a rock.

In the vivarium these lizards may be fed on the usual insects as well as on earthworms, but they do show a decided preference for spiders. In the case of Rykena & Nettmann (1978), they also ate raw meat cut into fine pieces. According to both authors, this species is more difficult to keep than other species of Green Lizard. Presumably they require a great variation in temperature and varying degrees of humidity. Perhaps it would be more expedient to keep them in an outdoor enclosure. According to Bech & Kaden (1991), this species may be kept in the same way as the Giant Emerald Lizard (*Lacerta trilineata*).

Lacerta strigata Eichwald, 1831
Caspian Green Lizard, Striped Green Lizard

Distribution: Eastern Caucasus and in Transcaucasia, especially in the northern Caucasus inclusive of Asia Minor through north and central Iran to southwest central Asia. These lizards live up to an altitude of 2,500 m.

Description: Striped Green Lizards reach a SVL of almost 11 cm. The tail is twice as long. The hindlegs, the hind quarters, and the tail are brown to grey-brown, but never green. The row of femoral pores runs down to the knee joint. Males are emerald green to olive green with scattered black flecks. The five rows of longitudinal stripes on the back and sides disappear almost completely from older males. Females are more strongly flecked and not such a bright green as the males. They are more-or-less brownish to greenish in colour. Females retain the five light longitudinal stripes. In some populations, males with blue throats are found. The underside is greenish white without flecks and has six rows of ventral scales. On the chocolate brown juveniles, the five yellowish white longitudinal stripes are especially vivid.

Habitat: These lizards live on rocky slopes with sparse vegetation but may also be found in meadows with individual bushes and in fields and forest glades. They can, furthermore, be found on the banks of standing or flowing water in forests, near ditches, and in vineyards.

Lifestyle, husbandry, and reproduction: According to the prevalent weather conditions, the animals emerge from hibernation in mid-March or early April and in northern areas some 2 weeks later. At an altitude of 2,000 m, the lizards only appear at the end of April or early May and go back to their winter quarters in mid-September. Thus the period of activity for these lizards is only 4 1/2 to 5 months. As opposed to this, the lowland species hibernate for around 8 months. This species requires only medium heat.

In spring and autumn the maximum temperature is only 22 to 25 °C. In summer it rises to over 26 °C, during which time the lizards are mainly active during the morning and early evening. Lizards that live near water will not hesitate to spring into the water and remain submerged when in danger.

As well as large insects, spiders, and ants, these powerful lizards will also eat small vertebrates such as smaller lizards and newborn mice. Sexual maturity is achieved at around 2 years of age. Females will usually lay their eggs at the end of May or in early June. Only very rarely are three clutches produced in one year. Each clutch contains between 6 and 11 eggs measuring around 15 to 18 × 8 to 10 mm. They take around 100 days to hatch in the vivarium. Newly hatched young have a SVL of 30.5 to 34.2 mm. These values may vary considerably within a clutch, possibly being a result of the food given to the gravid female. The young from some regions may be sexually mature at 21 to 22 months of age. They may begin mating after their second hibernation when they have a SVL of around 74 to 80 mm. The young from the second clutch of the year attain sexual maturity some 3 months later, and will lay their first clutch of eggs at the end of June or in early July. They will not lay a second clutch in their first breeding year. In Langerwerf's case (1980), the young that hatched in summer attained sexual maturity during the following December.

Lacerta trilineata Bedriaga, 1886
Giant Emerald Lizard

Distribution: The east, south, and west of the Balkan Peninsula and on some neighbouring islands. In the Caucasus region, in Asia Minor to Israel and Iran.

Description: With a SVL of over 16 cm and a tail 1 1/2 times as long, these lizards are significantly larger than the related European Green Lizard, *Lacerta viridis*. Several subspecies are known. European males are almost completely green. The juvenile pattern consists of three or five longitudinal stripes which gradually fade with

increasing age. The throat is yellow, whilst the sides of the neck are somewhat blueish. Females are similar to males, but they retain the stripes or they are a brownish colour with three to five noticeable longitudinal stripes. Juveniles are chocolate brown and also have three to five longitudinal stripes.
Habitat: Sunny, rocky hillsides with dense vegetation, bushy sand dunes, or immediately on the sea shore. In the Balkans up to an altitude of 1,000 m and in Armenia up to 2,000 m.
Lifestyle, husbandry, and reproduction: In the wild this species feeds on all manner of insects, snails and spiders, smaller lizards, small mammals, and nestling birds. Hibernation normally lasts from 4 to 6 months. Females are sexually mature in their second year after their second hibernation. Each clutch contains from 5 to 18 eggs.

Lacerta trilineata should be kept in pairs or in groups of one male with two or three females. Because of their size they should be kept in a very spacious vivarium that also has an aquatic section in which the lizards will frequently lie. Gravid females will often lie with their hindquarters in the water (Rogner, 1984). A mixture of sand and gravel is a suitable substrate. Cemented rocks will provide adequate hiding places. A spotlight should be installed above the highest rocks. This spotlight should produce a localised temperature of around 35 °C. Elsewhere in the vivarium the temperature during the day should be between 25 and 28 °C, falling at night to around 20 to 22 °C. Food takes the form of the usual insects as well as locusts, moths, flies, and snails.

These lizards should be hibernated for 6 to 8 weeks at 6 to 8 °C, before which they should be fasted for 2 weeks with the temperature being gradually reduced and the hours of daylight reduced to 9 to 10 hours. In the wild it is probable that males emerge from hibernation some 1 to 2 weeks before the females, since it is known from captive husbandry that males in the vivarium begin mating too late, causing females to produce infertile eggs, when both sexes are taken from hibernation simultaneously. It would appear wise therefore to take males from hibernation 2 weeks earlier than females.

Soon after sloughing the male assumes a rich green colour and follows the female relentlessly. After mating the females quickly increase in girth and the eggs may be clearly seen as lumps along the sides of the body. Eggs are usually laid in a moist place beneath a rock or are buried. In the vivarium these lizards should be given an annual cycle as similar as possible to that prevalent in the wild. A period in an outdoor enclosure with unfiltered sunlight will keep the lizards healthy and in good condition.

Pambov (1985) kept a pair of *Lacerta t. dubrogica* in a spacious vivarium. In April the lizards were placed with other species in an outdoor enclosure on a verandah. The enclosure had a ground area of 1.4 m². A heating pad connected to a thermostat tripped-in when the temperature fell below 18 °C. The outdoor enclosure may also be covered to prevent heat loss. During the day the temperature was around 24 °C, and at night fell to around 20 °C. The lizards were given a winter rest period at 15 to 18 °C during January and February, before which they had voluntarily stopped feeding. During summer they were fed on crickets, locusts, and earthworms; during winter on beef heart, liver, minced beef, aquarium fish, mealworms, and earthworms.

In June the female laid six eggs which quickly became mouldy. A second clutch of three hard and one soft-shelled egg were laid on August 25, 1983. They were incubated at a daytime temperature of 29 °C, falling at night to 24 °C, and 80% atmospheric humidity. Later the night temperature was increased to 26 °C and the humidity to around 85%. The young hatched after 82 to 84 days, measuring on average 6.5 cm. Only 5 weeks later the hatchlings had reached a length of 10 to 10.5 cm. One of the young hid itself away from the others and had a 6-week rest period after which it voluntarily reappeared. Meanwhile the other hatchlings had become much larger and stronger because they had remained active and continued feeding.

Lacerta "unisexualis" Darevsky, 1966
Virgin Lizard

Distribution: In small parts of Armenia and northeastern Turkey.
Description: With a total length of around 20 cm, these are medium-sized lizards. This somewhat squat, long-tailed lizard is reminiscent in colouring and patterning of Lacerta "uzzeli." It does however have additional blue eyespots behind the forelegs, whilst the dorsal markings often flow together to form a reticulated pattern.
Habitat: According to Franzen (1986), these lizards are found in relatively moist areas with fairly dense vegetation. They were however also found on roadside verges.
Lifestyle, husbandry, and reproduction: These lizards prefer to climb on rocks and cliff faces, living a lifestyle similar to that of other Rock Lizards, and should be kept in captivity in the same way and under the same conditions as described for other Rock Lizards. They should be hibernated for 5 to 6 months after which they will be much more ready to breed. During the activity period, the vivarium should be sprayed daily, this species having a great water requirement. This species reproduces parthenogenically.

Lacerta "uzzelli" Darevsky & Danielian, 1977

Distribution: In Turkey in the Kars Plateau and in neighbouring regions of Armenia.
Description: These delicate lizards reach a SVL of around 6 cm and have a tail some 11 cm long. Their ground colour is bronze-brown. There are two rows of dark spots along the centre of the back. At the front of the body there is a dark temporal band which is often broken by light spots. The belly is a whitish yellow.
Habitat: L. "uzzelli" is a species which is tied to cliffs and rocky areas and which often lives in large colonies. I was informed by Franzen that because of the climatic conditions in their natural habitat, this species hibernates for 5 to 6 months. These

Lacerta viridis

lizards require a great deal of water, making it necessary to spray their vivaria daily. In all other respects they should be cared for in the same way as other cliff-dwelling lizards and should be housed in a tall vivarium with a pile of rocks providing many crevices which they will use as refuges. This species reproduces parthenogenically.

Lacerta viridis (Laurenti, 1768)
Emerald Lizard, European Green Lizard

Distribution: West, central, and southern Europe, northern Asia Minor. In western France, on the upper reaches of the Rhine, in the southern Alps, Czechoslovakia, and in the southern European part of the former Soviet Union. In the south up to the northern parts of the Iberian Peninsula, Sicily and Greece. Also found on some Mediterranean Islands and some islands in the Aegean Sea.
Description: The European Green Lizard has a SVL of 13 cm with a tail twice as long. Males and many females are grass green to yellowish green on the upper side. The ground colour is often speckled with numerous small black spots. Some females, especially half-grown specimens, are uniform brown with black spots and have four narrow, light, longitudinal stripes. During the breeding season, males have a blue throat. The unflecked belly is a yellowish to whitish colour. The brown juveniles sometimes have light flecks on the flanks. The two to four, light, longitudinal stripes are one of the main distinguishing characteristics of this species.
Habitat: Sunny slopes with many bushes, the edges of fields, amongst hedgerows. Less frequently on heathland.

Lifestle, husbandry, and reproduction: European Green Lizards feed on all manner of arthropods, snails, and earthworms but will also eat small mammals and even the young of their own species. They should be cared for in the same way as *Lacerta trilineata*.

A clutch consists of between 5 and 21 eggs measuring around 14 to 15 × 8 to 9 mm. From these the young hatch after 7 to 9 weeks measuring 7.5 to 9 mm in length. Under favourable conditions, females may lay a second clutch of eggs. The young are sexually mature after their second hibernation.

Wittig (1986) kept this and other species in an outdoor enclosure with a ground area of 7 m^2. In this enclosure they proved to be more territorial than in an indoor vivarium and no young were found. It is possible that they had been eaten by other lizards. An outdoor enclosure with unfiltered sunlight, natural climatic variations, fresh air, and the most natural mixed diet of insects is obviously of advantage to this species, although their keeper is obviously able to study them easier in an indoor vivarium. Under ideal artificial climatic conditions, females may even lay up to six clutches in 1 year (Eichenberger, 1981), but that is simply too much. Equally undesirable are intergrades between various species. Ryken & Henke (1978) reported on matings between a male *Lacerta viridis* and a female *Lacerta agilis*. Two female intergrades resulted from this mating. One of the young was very similar in colour and markings to *Lacerta agilis*, whilst the other was an obvious mixture of the two parent animals.

Lacerta vivipara Jacquin, 1787
Viviparous Lizard

Distribution: In northern Europe from 70° north latitude southwards to northern Spain, northern Italy, and Bulgaria. In the east throughout almost the entire European part of the former Soviet Union, Siberia, the Altai, northern Mongolia to the Far East, and on the island of Sachalin.

Description: This attractive lizard reaches a maximum total length of 16 cm. Its very small head is barely set-off from the neck. The collar on the neck has pointed scales. The large scales along the back are keeled. The dorsal surface is chestnut brown, grey-brown, or grey, also occasionally partially bronze brown. Along the centre of the back there is a wide band that continues down the tail. This band consists of a more-or-less continuous raw of small dark spots. On the flanks there is a dark brown longitudinal stripe or row of spots. This is separated from the back by light and black side stripes. Males have a yellowish to orange underside that is speckled with black. Females have a white to mother-of-pearl underside with very pale or even no flecking. Totally melanistic or totally green specimens are also known.

Habitat: Heathland, the edges of forests, light forests, moist meadows, young plantations, moors, sand dunes with vegetation, and old abandoned quarries.

Lacerta vivipara

Lifestyle, husbandry, and reproduction: In the wild, the Viviparous Lizard feeds on earthworms, centipedes, spiders, and insects and their larvae. In Central Europe, hibernation lasts from around October until late February or early March. Mating takes place from April to June. Females are gravid for 3 months, giving birth to 3 to 10 young from July to October. The young immediately break free from the amniotic sac. The transition from being egg layers to becoming live bearers is part of the evolutionary history of these lizards. In Spain there are populations of *Lacerta vivipara* that produce parchment-like eggs after only around half the normal embryonal development time. The young hatch from these eggs at the end of the normal development period (Osenegg, 1991). In the most northerly part of their distribution range, as an adaptation to the short summer, the lizards mate in autumn producing young the following spring. These lizards, which are not sensitive to climatic variations, may be kept in an outdoor enclosure throughout the year. Here they will also readily reproduce. If the enclosure is planted as heathland, with lawn, or made to imitate the edge of a forest, this will attract many insects that the lizards will happily devour in great numbers. Additional food should also be given to ensure that the animals are well nourished before hibernation. Earthenware pipes penetrating at an angle some 60 to 80 cm into the ground will provide ideal winter quarters. The entrances to these earthenware pipes should be concealed by tree roots. The pipes may be partially filled with sphagnum moss which gives excellent protection against frost. In most European countries, it is forbidden to catch wild lizards and only captive-bred animals may be kept. Captive-bred young must not be released into the wild.

Meroles Gray, 1838
Sand Lizards

This genus contains six or seven species that live in the western dry regions of South Africa. Apart from one species, all are endemic. *Meroles* species are distinguished by fine granular scales on the back and a well-developed collar. The supraoculars do not reach the lips. The eyelid is not transparent. The feet are well-developed and have fringed toes. Femoral pores are always present. According to Arnold (1989), the once monotypic *Aporosaurus anchietae* should be reassigned to the genus *Meroles*.

Meroles anchietae (Bocage, 1867)
Dunes Lizard

Distribution: Endemic to the Namib Desert from the Lüderitz Bight to southern Angola.
Description: This species usually reaches a length of 10 to 11 cm, with the maximum length being around 12.5 cm. They may be easily recognised by their flat, angular snout. In comparison to other *Meroles* species, *Meroles anchietae* does not have a collar or femoral pores. The hindlegs are long and point outwards. There is a fringe on each of the long toes. This fringe is composed of enlarged scales. The tail is very thick at the root, gradually tapering towards the tip. The sand-coloured upper side has a reticulated pattern above the flanks. The pale head has a silvery sheen. Sometimes there is a broken stripe running from the back of the head to the tail joint. Occasionally there are black transverse bands on the tail. The underside is white.
Habitat: Sand dunes.
Lifestyle, husbandry, and reproduction: These diurnal ground dwellers are well adapted to life in the desert and can run quickly over the sand. Although they can withstand temperatures up to 44 °C, they may overheat on the hot sand. To prevent this they perform an interesting "thermal dance" during which they raise the tail and two feet at once from the ground. When in danger, they use swimming movements to bury themselves quickly in the sand. If they are unable to escape, they will rear-up the body and spring towards their attacker attempting to bite in self defence. Especially during the breeding season this species will claim a territory that is vigourously defended by the males. Their food consists mainly of hard-shelled beetles and other insects. During the dry season they will also eat seeds. Liquids are

Meroles anchietae

obtained from their food. Females lay one or two eggs in a chamber that they dig into firm sand. There is no rigid incubation period, but most eggs hatch between December and March.

In the vivarium this species should be kept in the same way as the *Eremias* species, but without hibernation. During the day the temperature should be around 30 °C, falling lower at night. Beneath a spotlight there should be a localised temperature of 40 °C. In the vivarium this species will greedily devour all manner of hard-shelled beetles and insects.

Meroles cuneirostris (Strauch, 1867)
Desert Lizard

Distribution: Limited to the southern Namib Desert from Richtersveld to the Walvis Bight.
Description: This 13 to 16 cm lizard has a flat, wedge-shaped head and significant fringes on the toes. The ear openings are covered by a fold of skin. The nasal scales are swollen and do not meet at the top. At the centre of the body there are 90 to 110

Meroles cuneirostris

rows of scales. There are 18 to 24 femoral pores. The back is grey, reddish brown, or sand-coloured, sometimes with a pattern of dark flecks, which in males may form a reticulated pattern. A longitudinal stripe that is light above and dark below separates the back from the flanks. This stripe runs from the eye to the tail. The folds of skin on the sides and the upper labial scales may be yellowish orange in colour. The underside is white.

Habitat: Low sand dunes and sandy areas with sparse vegetation.

Lifestyle, husbandry, and reproduction: The variety of food available depends upon the season. In summer Desert Lizards eat mainly beetles, whilst in winter they eat termites and other small insects. As ground dwellers, they hunt mainly in the morning and early evening when it is cooler. The lizards will often sit by an ant track, picking off the insects that the ants have caught. When in danger, they burrow into the loose sand where they also spend the night. There is no distinctive breeding season. Females lay two to four eggs up to 60 cm deep in the soft sand. Hatchlings are around 56 mm long. In contrast to other *Meroles* species, Mayer & Richter (1990) describe these lizards as "ideal vivarium Lacertids."

For an individual pair, a vivarium of 40 × 40 × 50 cm is suitable. It should contain a substrate of lime-free sand, the surface of which should not be allowed to form a crust. The lower reaches of one section of the substrate should be kept constantly moist. Some rocks, climbing branches, and bundles of dry grass complete the furnishings. Although the vivarium should always contain a shallow bowl of fresh drinking water, the vivarium only needs to be sprayed very occasionally. Incandescent lamps are usually used as a heat source. HQL lamps are especially suitable for this purpose. Ultraviolet strip lights should also be fitted in the vivarium. These should be in operation for a short time twice or thrice weekly. Simple fluorescent tubes and ultraviolet are however quite adequate. A multivitamin preparation and calcium lactate should be given every 14 days. Additional calcium supplies are also advisable. These lizards may be fed on the usual insects. Meadow sweepings have proved to be invaluable since these usually contain hard-shelled beetles, bugs, and some ants.

It is quite easy to reverse the seasons. During the European summer additional heating is left quite low, but increased in November until the daytime temperature is around 30 to 32 °C, increased locally beneath a spotlight to 40 °C. At night the temperature should fall to around 20 °C. When the heating is increased in autumn, the hours of lighting are also increased to 16 hours per day. Before that, 10 to 12 hours is sufficient.

When the lizards have adapted to this change, they will start mating. The date when the eggs are laid will only vary slightly from the time they would normally be laid in the wild. Females lay between two and five eggs measuring around 9 to 10 × 5 to 6 mm. In the wild, the young hatch after 70 to 90 days with a SVL of 26 to 29 mm. In captivity, these lizards quickly become tame. As soon as the keeper approaches the vivarium, they come to the front to "beg" and will not hesitate to take food from the fingers. In such situations food should be given sparingly since the lizards will quickly become obese. In the vivarium these lizards will frequently lie on their backs with outstretched legs beneath the spotlight. They will then suddenly right themselves and disappear into the sand.

Meroles knoxii (Milne-Edwards, 1829)
Knox's Burrowing Lizard

Distribution: Along the western Cape coast from the peninsula to the coastal region of southern Namibia and inland from Majesfontein to Tanquwa Karoo (endemic).
Description: These desert lizards reach a length of 15 to 20 cm, maximum 23 cm, and have a rounded snout without any sharp edges. Specimens from the Cape Peninsula remain smaller, reaching a length of only around 18 cm. The overlapped ear openings are clearly visible. The supranasals do not touch one another. Dorsal scales may be smooth, as in some northern populations, or lightly keeled. At the centre of the body there are 54 to 78 rows of scales. The ventral scales form 10 to 12 longitudinal rows. The femoral pores are well-developed. Along each of the toes there is a small fringe. The dark brown to black juveniles have five white stripes and white speckling. Adults are reddish brown. Specimens from the coastal region are paler with a dark dorsal stripe interrupted by many white-edged black spots. The

Meroles suborbitalis

flanks are grey with yellowish or brownish flecks or circles. There may also be pale flecks on the legs. The underside is cream to white or blue-grey. Sexually mature males have vivid flecks on the throat and anal regions.
Habitat: Coastal sand dunes and in areas overgrown by succulents.
Lifestyle, husbandry, and reproduction: These very active, diurnal lizards scurry around the sandy areas between bushes searching for all manner of insects, especially beetles and flies. As refuges, they dig burrows in the firm ground at the base of succulents. The size of the clutch varies from female to female. Smaller specimens from the Cape Peninsula lay only two or three eggs, whilst larger lizards from Namaqualand lay up to six. Mayor & Richter (1990) found *M. knoxii* to be very delicate. They were always extremely anxious and intolerant of one another.

Meroles suborbitalis (Peters, 1869)

Distribution: Central Karoo to Namaqualand and southern Namibia. An isolated population lives in the central Namib Desert.
Description: These medium-sized lizards reach a total length of 16 to 18 cm, at most 21 cm. They have a rounded snout without sharp edges on the upper lip. Their overlapped ear openings are visible. The supranasals do not touch one another. There are 60 to 75 rows of scales at the centre of the body, whilst on the underside there are 12 to 14 longitudinal rows of scales. This species has femoral pores and

slightly fringed toes. On a pale yellowish white background juveniles have four black stripes and a black reticulated pattern with large pale flecks on the legs. In southern representatives of the species, the stripes change to light brown circles with increasing age. The grey-brown back is covered in large pale flecks, sometimes arranged in rows. In the Namib Desert, adults are slate-coloured with a pinkish sheen and irregular rows of dark flecks with pale edges. The underside is a creamish white to blueish colour. Males sometimes have a yellow throat.

Habitat: Usually in flat, sandy areas with low bushes and grasses, less frequently on sand dunes.

Lifestyle, husbandry, and reproduction: These excellent hunters live on sandy ground or areas of fine, smooth gravel with many bushes. They hunt from the cover of low vegetation, preying on beetles, crickets, and termites. It would appear that in the Namib Desert, this species reproduces throughout the year. Females lay between three and seven eggs. In the Kalahari Desert these lizards mate in early winter. Females lay up to two clutches, each containing four to eight eggs. The young grow very slowly, taking 3 years to reach sexual maturity.

According to Mayer & Richter (1990), this species is very sensitive and delicate in captivity. It is also very intolerant of others of the same species. They require a very spacious vivarium with a deep layer of sand and several groups of plants. They should be kept in the same way as *Meroles cuneirostris*.

Nucras Gray, 1838
Blunt-Headed Lizard

The members of this small African genus have a rounded snout, a cylindrical body, and an extremely long tail. The subocular scale touches the labial scales. The lower eyelid has scales. The nostrils lie between the second and third nasal scales and are distinctly separated from the labial scales. The collar is very prominent. The small, smooth dorsal scales do not overlap. On the toes there are small lamellae, but no fringes. Femoral pores are present.

Because they have a very secretive and concealed lifestyle, these nocturnally active lizards are often overlooked. They hunt in the early hours of the morning or in the evening. They are often seen eating termites. The long tail serves as a fat reservoir. All members of this genus are live-bearing species. They are limited to dry temperate savannahs with dry sandy ground or "fynbos," i.e., areas with a Mediterranean climate and grassy hillsides in the mountains. There are seven species known from the south and east of Africa. The species are mainly distinguished by their colouring and less by the number of scales.

Nucras intertexta (Smith, 1838)

Distribution: Southern Africa but not down to the Cape.
Description: This species may reach a length of up to 28 cm, but a total length of 22

to 26 cm is more usual. They have a speckled back and an orange-brown tail. Below the forelegs there are four to seven enlarged plates. At the centre of the body there are 34 to 56 rows of scales. The back is light brown to red-brown, speckled with numerous pale flecks, which may have dark edging or which may form irregular transverse bands. Sometimes there is a pale central stripe. The underside is creamish white with dark flecks on the outer rows of scales. Juveniles are darker on the back and have rows of vivid cream flecks. The sides of the neck have pale yellow bands and the tail is coral red.
Habitat: Dry savannahs, usually on Kalahari sand.
Lifestyle, husbandry, and reproduction: As for *Meroles* species.

Ophisops Menetries, 1832
Snake-Eyed Lizard

The distribution range of the Snake-Eyed Lizard stretches from North Africa through near Asia to India. These lizards live in dry, stony areas, usually with sparse vegetation. The body is long and rounded. The tail is around twice the SVL. In this species the dorsal scales are large, overlapped, and heavily keeled. There are also keeled lamellae on the toes. The collar is poorly developed or even incomplete in the centre. The eyelids are fused together with the lower eyelid having a transparent "window." These lizards can, however, close their eyes by pressing the lower eyelid into a fold. This is a mainly brown to olive green lizard that often has light side stripes and dark flecks. The tail is usually a reddish colour.

Ophisops elegans Menetries, 1832
European Snake-Eyed Lizard

Distribution: In Europe: European Turkey, southern Bulgaria, northeastern Greece, some islands of the northern Aegean Sea, eastern Caucasus. Also in Asia Minor and Cyprus to Iran. Furthermore, around the Mediterranean Sea to Libya.
Description: These small attractive lizards reach a length of 16 to 18 cm of which one-third is SVL. The ground colour is brown, which may change to green on the flanks. Along each side there are two narrow light stripes that are especially noticeable in juveniles. Between these stripes and the back there are black flecks, dashes, and spots that are usually arranged in rows. The underside is uniform white to grey-green but may occasionally be a reddish or yellowish colour. In the breeding season, males have a pale blue throat and blue flecks over the shoulders.
Habitat: Dry, steppelike land with sparse vegetation. Also found on cultivated land, in parks and in light forests.
Lifestyle, husbandry, and reproduction: In the wild, these lizards feed on all manner of arthropods, being especially fond of ants. Males are extremely aggressive towards one another, especially during the breeding season, and will fight very vigourously, during which time they indulge in ritualised head bobbing. The first eggs are laid at

Ophisops elegans

the end of April, after which a second and sometimes even a third clutch follows. Each clutch consists of three to six eggs measuring around 10.5 × 5.5 mm. On hatching the young have a SVL of 21 to 24 mm and have four significant white longitudinal stripes. These lizards may be kept in the same way as *Acanthodactylus boskiana*. Mantel (1986) kept two male and one female *O. elegans ehrenbergii* from Israel in a vivarium measuring 70 × 70 × 40 cm with a 5 cm layer of sand as substrate. Bunches of dry grass and some flat rocks completed the furnishings. Within 1 year the female laid three clutches, each of two eggs, which hatched after 50 days at 29 °C. The young should be reared on small insects that have been dusted with calcium and a vitamin/mineral preparation.

Ophisops occidentalis Boulenger, 1887
Western Snake-Eyed Lizard

Distribution: Morocco to Libya.
Description: This lizard reaches a total length of around 15 cm and is similar to the European Snake-Eyed Lizard. At the centre of the body there are 26 to 30 rows of scales; in *O. elegans*, 30 to 40.

Ophisops occidentalis

Habitat: Steppes and cultivated land with sparse vegetation.
Lifestyle, husbandry, and reproduction: See *Ophisops elegans*.

Pedioplanis (Duméril & Bibron, 1839)
South African Racerunner

Small lizards with a cylindrical body and long tail. The subocular scale touches the lips, whilst the lower eyelid may have scales or a transparent "window." There is a well-developed collar. The dorsal scales are usually small and smooth and do not overlap. The long toes do not have fringes, but there are keeled lamellae on their underside. This collared lizard has femoral pores.

Pedioplanis breviceps (Sternfeld, 1911)
Short-Headed Racerunner

Distribution: Western areas of the central and northern Namib Desert of northwest Namibia.
Description: These lizards usually reach a total length of 12 to 14 cm, rarely 16 cm. The head is short with scales on the lower eyelid. The eardrum is not enlarged.

There are 12 rows of scales on the underside. The lower reaches of the limbs have small, unkeeled scales. Males and females differ in colour. Juveniles and adult females have five distinct black stripes along the back. In young males, the stripes gradually disappear so that adult males have a uniform light brown back that has many grey flecks, occasionally arranged in rows. The underside is white. The light brown tail has thin dark brown bands that change to a reddish colour nearer the tip.
Habitat: Desert.
Lifestyle, husbandry, and reproduction: These very active lizards hunt their prey in open, sandy areas and dry river beds, feeding on insects that they usually hunt in the early hours of the evening. In the wild, eggs develop from December onwards. It is only between March and May however that the females lay a clutch of two to four soft-shelled eggs. The young hatch around 2 months later, measuring from 50 to 80 mm in length.

The Short-Headed Racerunner requires a spacious vivarium with a sand substrate around 10 to 20 cm deep. Although the lizards are rarely seen during the day, the vivarium should be well lighted and fitted with a spotlight that produces a localised temperature of 35 to 40 °C. The lower reaches of the sandy substrate should be kept slightly moist. The only furnishing required is a dry tree root. The lizards will often only appear after the lights have been switched off. It is at this time that they should be fed and the vivarium lightly sprayed.

Pedioplanis lineoocellata (Duméril & Bibron, 1839)
Flecked and Striped Racerunner

Distribution: The western half of South Africa, but not in sandy regions such as the Namib or Kalahari Deserts.
Description: This lizard reaches a length of between 13 and 17 cm, rarely 19 cm. It has a slightly compressed head and body. The eardrum and temporal scales are not enlarged. The lower eyelid has a "window" composed of two black-edged, transparent scales. In front of the ear opening there are three or four "lobes." These lizards vary greatly in colour. The back has ochre to grey-brown or reddish brown tones, usually with two to four rows of small pale flecks that run together to form a continuous or broken stripe. Along the flanks there are four to seven rows of large, pale blueish-white, sometimes dark-bordered, flecks. Animals from the Lüderitz Bight area are dark grey, usually with four pale dorsal stripes, but without flecks on the flanks. It is possible that these animals are a subspecies *(P. l. mocellata)*. Some specimens from Namaqualand are distinguished by their vivid colouring and pairs of black dorsal stripes. The hindlegs are flecked. There are black and pale flecks on the tail, whilst the underside is white. Males sometimes have a grey-blue throat.
Habitat: Varies considerably. Fields, bushy valleys, dry and moist savannahs.
Lifestyle, husbandry, and reproduction: This species prefers flat, stony areas in which they may hunt insects under cover. Their food varies according to the seasons.

In winter they feed mainly on termites, whilst in summer they eat beetles and crickets. These Racerunners are diurnally active, even on warm winter days. They seek refuge in hollows beneath rocks or in rock crevices. When in danger they leap from a rock, making them extremely difficult to catch. It is only when running on open land that they tire quickly and become easy to catch.

In November, females lay between four and eight eggs in short tunnels that they dig themselves in the moist ground near rock crevices. The 50 to 55 mm young hatch after 70 to 80 days and are sexually mature at around 15 to 18 months of age. Unfortunately, this species only lives for 4 or 5 years. They should be kept in the same way as *Pedioplanus breviceps*.

Pedioplanis namaquensis (Duméril & Bibron, 1839)
Namaqua Racerunner

Distribution: East of the Cape through Karoo, Namaqualand, Namibia, and Botswana to southern Angola.
Description: These 14 to 17 cm (maximum, 18 cm) lizards are distinguished by their fairly long tail and enlarged eardrum scale. The lower eyelids are semi-transparent. In front of the preoculars, there are three to eight granular scales. The ear openings do not have "lobes." At the centre of the body, there are 47 to 65 rows of scales with 12 to 14 rows of ventral scales. The black young have four narrow white stripes. The legs are brown with white flecks, whilst the underside is white and the tail a pinkish brown. Animals that live on the dark ground of the Karoo Region frequently retain the juvenile colouring. In specimens from western regions, the dorsal stripes often change to light brown and the flanks have irregular bands.
Habitat: Dry savannahs and semi-deserts.
Lifestyle, husbandry, and reproduction: During the day these lizards scurry around the sandy and gravely areas searching for insects amongst the sparse vegetation. Their food consists of small insects. They spend the winter in burrows that they dig in the sand at the base of a bush. In November the females lay between three and five eggs. This species should be kept in the same way as *P. breviceps*.

Podarcis Wagler, 1830
Wall Lizards

The main distribution range of these small- to medium-sized lizards stretches from southern Europe to North Africa and Asia Minor. They differ from members of the genus *Lacerta* by the construction of the skull, the tail vertebrae, and the hemipenes, but are very similar to Lacerta in appearance. Their dorsal scales are often lightly keeled; the collar is easily recognisable and is slightly toothed.

Amongst these lizards, brown and green are the predominant colours. The markings usually consist of light, and sometimes dark, longitudinal stripes and flecks. Fe-

Lizards

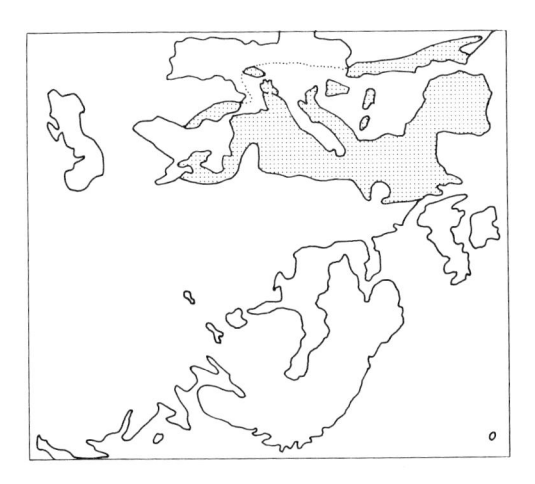

Distribution of *Podarcis*

males frequently have distinctly defined longitudinal stripes whilst in males the pattern usually consists of spots and whorls. In some species, e.g., *P. meliselensis*, *P. taurica*, and *P. sicula*, specimens without markings occasionally occur. These are designated "concolour" forms. The melanistic island forms of *P. lilfordi*, *P. sicula*, and *P. pityusensis* are worthy of special note. Even animals from closely neighbouring islands are so divergent in colour that they are classed as individual subspecies. These lizards require a vivarium at least 60 cm long and are usually kept in pairs or small breeding groups of one male and two females. *Podarcis* species feed on all manner of arthropods and will also lick on sweet fruit. The vivarium must contain a shallow drinking bowl and must be fitted with a spotlight. The daytime temperature should be between 25 and 30 °C.

Podarcis bocagai Seoane, 1884
Bocage's Wall Lizard

Distribution: Northwest Spain and Portugal.
Description: These sturdily built lizards reach a SVL of up to 7 cm and have a tail some 117 mm long. They are very similar to *P. hispanica* which occurs in the same areas, but males often have a green back whilst the stripes on the females are much paler. Even males with a brown dorsal surface have at least a green stripe along each side of the back. These Wall Lizards have a yellow, orange, salmon pink, or sometimes white underside, which, like the throat, is usually heavily flecked.
Habitat: From sea level up to an altitude of 1,500 m. Stony slopes, rocky areas, roadsides, fallen trees, and areas of sparse vegetation.
Lifestyle, husbandry, and reproduction: Bocage's Wall Lizard is active from March

until late November. Males emerge from hibernation earlier than females. These lizards hibernate individually as well as in pairs or in groups of three. They overwinter in tunnels that they dig themselves, or deep in the centre of rock piles. The main period of activity is between 1600 and 1700 h. They feed on all manner of insects. Mating begins in spring when the females emerge from hibernation. In the wild, mating takes place during March and April. During May and June the females lay from one to five eggs, the average size of which is 10.6 × 6.06 mm. It is thought that these lizards reach sexual maturity at around 2 years of age.

Podarcis dugesii (Milne-Edwards, 1829)
Madeira Wall Lizard

Distribution: The island groups around Madeira, also on the desert islands. On Porto Santo and neighbouring islands and on the Azores. The population on the Selvagens was possibly introduced by humans.
Description: With a total length of over 23 cm, these Wall Lizards are some of the largest and most powerful of the genus. The tail is usually slightly less than double the SVL. The relatively pointed head appears swollen around the cheeks, especially in males. The body is usually brown or grey-green to black with yellow or greenish spots or speckles. The light longitudinal stripes, which are dark-edged, run from the eye to the root of the tail. A dark stripe with lighter speckling runs along the flanks. As the animals grow older, the stripes gradually disappear or become very pale and barely noticeable. There are sometimes black spots on the cream or yellowish underside. Completely melanistic specimens occur frequently; their underside often has a blueish tone.
Habitat: It is thought that earlier these lizards inhabited stony, rocky areas with macchia-like vegetation. Nowadays they are found in the most divergent habitats, even on the coast, down to the waterline. Inland, they are frequently found near streams.
Lifestyle, husbandry, and reproduction: In the wild, this species feeds mainly on ants, although in the vivarium they may be given the usual variety of insects. In some populations, these lizards will also eat fruit and other sweet matter, or they will at least lick such matter.

Males are very aggressive. This species mates in the usual way. Exact information on their breeding habits is only available from vivarium observations, which report that mating begins in April, around 3 to 4 weeks after hibernation. From May to August the females laid two, and on one occasion three, clutches of eggs. The interval between each clutch varied from 5 to 8 weeks. At temperatures from 28 to 30 °C during the day and 20 to 25 °C at night, the eggs took between 80 and 91 days to hatch, with the young measuring from 85 to 93 mm. These animals are sexually mature at around 15 months, irrespective of whether they have hibernated (Richter, 1986).

Podarcis erhardii (Bedriaga, 1876)
Aegean Wall Lizard

Distribution: Southern Balkan Peninsula, Crete, and some of the Aegean islands. On the islands they are usually the only species of lizard found.
Description: There are numerous subspecies of this lizard. On the mainland they are reasonably uniform in colour. They generally reach a SVL of up to 7 cm with the tail being twice as long. In most cases, these lizards are a brown to sand colour with a bronze to copper metallic sheen. A dorsal stripe is sometimes present. The stripes along the sides of the back are dark and much wider, whilst the sides themselves often have a reticulated pattern. Females usually have very conspicuous side stripes. Along the outer rows of ventral scales, there are usually blue spots which may occasionally also extend to the flanks. Males usually have an orange-red underside, whilst that of females is a mother-of-pearl colour. Dark spots are often found on the throat. Some island populations vary considerably in colour. As well as a brown background colour, there are also grey, green, and very dark brown specimens. The dark patterning may be reticulated, but in some cases may be completely absent. At present, 7 mainland and 24 island subspecies are recognised, but herpetologists are unable to agree on these figures.
Habitat: Dry rocky areas with bushy vegetation.
Lifestyle, husbandry, and reproduction: In the wild, these lizards eat all manner of insects, butterflies, and flies, which they catch near the nesting sites of Eleonore Falcons. These falcons nest on the ground, but near the nest do not eat these lizards which would otherwise form part of their normal diet. The food remnants around the nest attract countless flies, which in turn fall prey to the lizards. After the young falcons have left the nest, the lizards again disperse over a wide area.

Males fight bitterly amongst themselves. They claim and defend a distinct territory that is normally delimited by small mounds of excrement. Mating takes place in March or April with the females laying two or sometimes three clutches of eggs in mid-July. The number of eggs in each clutch varies from one to four, each measuring around 13 to 17 × 8 to 10 mm. The size of the eggs varies according to the subspecies. The 8 to 10 cm young hatch in September and are very similar to their parents in appearance. Even in the vivarium, males are extremely intolerant of one another.

Podarcis hispanica (Steindachner, 1870)
Iberian Wall Lizard

Distribution: Iberian Peninsula, southeast France and the Columbretes Islands off Castellon. There are individual populations in northwest Africa.
Description: With a SVL of barely 6.5 cm, the Iberian Wall Lizard is one of the smaller species. The tail is around double the SVL. The upper side of the body is mainly brown and grey with the underside being white, yellow, pale red, or even

Podarcis erhardii (above), Podarcis hispanica (below)

deep red. There are numerous small spots on the throat. Females usually have five to seven dark longitudinal bands below which there are four light lines. The males, which are mainly spotted, flecked, or reticulated, often have blue flecks along the flanks. If a dorsal stripe is present, it is usually much paler than the dark stripes along the sides of the back. Specimens completely without markings are mainly found in the east of the distribution range. Even within a given area, the colouring of these lizards can vary considerably. It can therefore be difficult to distinguish them from other species occurring in the same area.

Habitat: These lizards are excellent climbers on perpendicular surfaces such as cliff faces, scree-covered hillsides, tree trunks, and roadside escarpments. They do, however, show a distinct preference for old, dry, limestone walls.

Lifestyle, husbandry, and reproduction: A large number of individuals usually live together, even in a very small area. Males claim only a very small territory but are not especially aggressive towards one another. They feed on all manner of small insects. In May, females lay two, or, according to the geographic location, at most three, clutches of one to five eggs. On average, these eggs measure 6 × 11 mm from which the young hatch after 2 months measuring around 5 cm in length. Under captive conditions, mating took place from late February until early April with a clutch of eggs being laid on March 9. Courtship involves the male following the female for some time before biting her on the tail. A female willing to mate will indicate this by kicking or "paddling" with the hindlegs. The actual copulation lasts from 5.5 to 50 minutes. Verbeer (1972) saw that during mating the male uses the hindlegs to hold the female in position. The eggs measured 11.3 to 12 mm × 6 to 7 mm. The young hatched after 63 days with a total length of 51 mm of which 23 mm was SVL. Algerian specimens of this species are somewhat larger. Gravid females with a SVL of 39 mm have been found.

Podarcis lilfordi (Günther, 1874)
Balearic Wall Lizard

Distribution: Balearic Islands (endemic).

Description: With a SVL of around 8 cm and a total length of 18 to 20 cm, these lizards are one of the larger species. The body is stocky and powerful whilst the head and tail are quite stout. The colour varies from brown to green or olive. Usually between the light stripes on the sides of the body there are three dark, partially broken bands. The flanks normally have a reticulated pattern with some blue flecks. The white, yellow, or reddish belly often has dark patterning. Many populations tend to dark and even black colouration. Such specimens have a dark brown to black upper side and a deep blue underside. The geographic isolation of the animals on many smaller islands leads to numerous variations in colour. For this reason, 13 subspecies are recognised.

Habitat: These lizards occur in huge numbers on the small islands and rocky outcrops around the main islands of Majorca and Menorca. On both main islands there are only very small populations, such as that on the city walls of Palma on Majorca.

Lifestyle, husbandry, and reproduction: These lizards feed mainly on insects and spiders but will also eat snails and woodlice. The amount of plant material they also eat is worthy of mention. They will even eat leftovers of human food, and, when food is in short supply, they will not hesitate to resort to cannibalism when they will readily eat their own eggs and young. Balearic Wall Lizards are not timid in captivity and quickly become quite tame. They are, however, aggressive towards one another, making a very spacious vivarium essential. If fighting continues, the opponents must be separated. During warmer months of the year, it is best to keep these lizards in an outdoor enclosure that is partially covered with glass or rigid transparent plastic sheeting. A clutch usually consists of two or three eggs; very rarely are four produced.

On March 21, 1978, Langerwerf (quoted by Salvador, 1986) removed a clutch of two eggs from a breeding vivarium. These eggs had probably been laid several days earlier. For the first 2 weeks the eggs were incubated at 31 °C after which the incubation temperature was reduced to 27 to 28 °C. The young hatched between July 8 and 15. Langerwerf successfully incubated a second clutch of four eggs at 27 °C. The young are easily reared on small insects dusted with vitaminised calcium. The vivarium should be sprayed daily. A multivitamin preparation should be added to the drinking water twice weekly.

Podarcis melisellensis (Braun, 1877)
Adriatic Wall Lizard

Distribution: Adriatic coastal regions of northern Italy, Yugoslavia, and northern Albania up to an altitude of 1,300 m above sea level. Also found on numerous small, offshore islands.
Description: In these attractive, short-headed lizards, the SVL is usually less than 6.5 cm with the tail being twice as long. The dorsal surface is brown or green. There is a light stripe along each side of the back. These enclose a broader, darker stripe that is broken by light flecks. Along the flanks there are usually rows of light or dark flecks with blue scales along the sides of the belly. Males have a blue shoulder fleck. Specimens without markings are quite common. Females have a white underside, whilst that of males is vivid yellow, orange, or red.
Habitat: Chalky ground or rocky hillsides with sparse vegetation.
Lifestyle, husbandry, and reproduction: Males are usually very aggressive towards one another. It is thus only possible to keep this species in pairs or small breeding groups of one male with two or three females. To prevent intergrades from being produced, closely related subspecies should not be kept together. Many lizard breeders begin by initially housing two males together, with females being introduced later; the stronger of the two males will then mate with the females. The subservient male should then be removed, as should the females when it becomes obvious that they are gravid. This prevents them from being molested in any way by the other lizards. Stacked rocks and climbing branches allow the lizards to escape from one another should a conflict occur. One corner of the sand/peat substrate should be

kept constantly moist so that eggs that have been laid in the vivarium without being seen may be allowed to develop. The temperature in the vivarium should be between 25 and 30 °C. To be able to keep a better check on the eggs however, they should be placed in an incubator. A few days at cooler room temperature will not harm these lizards. Instead, it will give them a welcome rest period. After fasting them, the lizards may be placed in suitable boxes, filled with peat, and placed in a cool cellar at 4 to 5 °C for hibernation. To keep the peat moist the boxes should be well covered. Fresh air is allowed to enter during the weekly check of the lizards. If the peat or sphagnum moss in the boxes becomes too dry it should be lightly sprayed. Mating begins immediately after hibernation, with the eggs being produced around 3 weeks later. The four to five eggs are laid in a hole that the female digs herself. After their first hibernation, young females usually lay only two eggs. The eggs are always well covered and difficult to find. Some females may lay up to five clutches of eggs in 1 year. The shortest time known between clutches is 16 days. At a mean temperature of 27 °C the young hatch after around 39 days and should be reared separately from the adults.

Podarcis muralis (Laurenti, 1768)
Common Wall Lizard

Distribution: Central and southern Europe.
Description: These Wall Lizards reach a total length of 20 to 25 cm of which 7.5 cm is SVL. The body is fairly compressed and, according to population, can vary considerably in colour and pattern. Many specimens have a black-flecked pattern on a brown background. This flecking may often form a reticulated pattern. There are frequently also two light longitudinal stripes. In such cases the central stripe is much more vivid than the stripes along the sides of the back. These side stripes may even be completely absent. There is a band of dark flecks along the flanks. These are sometimes enclosed by light stripes. The scales along the sides of the belly are often blue. Females are usually more brightly coloured and more vividly marked than males. The underside of the Common Wall Lizard may be white, yellow, orange, or even red, sometimes rust-red with dark spots, especially on the throat. The underside of males is usually more colourful than that of the females. Around 20 subspecies are recognised.
Habitat: Dry, stony and sunny areas, walls around fields, rock piles, light forests, moist banks of mountain streams up to an altitude of 2,000 m. Often found amongst ruins and on dry stone walls.
Lifestyle, husbandry, and reproduction: Common Wall Lizards eat all manner of insects as well as snails and earthworms. They very rarely eat fruit. Males defend their territory very vigorously against intruders of the same species. According to the prevalent climatic condition, hibernation may last for several months or only for

several weeks. The mating which follows lasts only 30 to 120 seconds. Mating is preceded by a "mating march" which serves to coordinate the movement of the two partners and helps the male to orientate himself to the movements of the female. For females, this "mating march" forms an essential part of the courtship.

In May or June the females lay their first clutch of 2 to 12 eggs, each measuring around 10 to 12 × 5 to 7 mm. Females may lay up to three clutches in 1 year. Those living at higher altitudes however lay only one clutch. In the wild, the development of the eggs takes from 2 to 3 months before the young hatch measuring around 6 cm in length. In more favourable locations the young may even hatch at the end of July. These lizards are sexually mature at around 2 years of age.

At a constant temperature of 26.7 °C, incubation takes less than 7 weeks, but at a daytime temperature of 18.3 °C, falling at night to 12.8 °C, incubation may take more than 5 months (Cooper, 1958, 1965). Even within a single clutch of eggs, the incubation time may vary considerably. After cutting through the shell, the young usually remain inside the egg for a few hours or even up to two days.

In the vivarium the Common Wall Lizard should be kept in breeding groups of one male with two to four females. The vivarium should have a floor area of around 80 × 40 cm and should contain stacked rocks and a spotlight. During the day these lizards require a temperature of 28 °C, falling at night to 18 to 20 °C. Hibernation for 1 to 2 months is essential for breeding. The young should be reared separately from the adults.

Podarcis peloponnesiaca (Bibron & Bory, 1833)
Peloponnese Wall Lizard

Distribution: Limited to the Peloponnese where it is endemic.
Description: A robustly built lizard which, with a SVL of over 8 cm, is one of the larger species. The tail is around 16 cm long. Specimens with a total length of over 25 cm are known. Along the back the scales are usually very smooth but occasionally lightly keeled. The ground colour on the back is usually brown but in males may be green with three or four insignificant longitudinal stripes. Between these stripes there are bands of black spots, which, especially on the sides, may form a partially reticulated pattern. A certain characteristic for identification is the vivid blue shoulder flecks which sometimes continue along the flanks. The belly is also frequently a blue colour.

On a brown background, females have four light contrasting stripes. The two outer stripes may run together at the front of the body. Females occasionally also have the blue shoulder flecks. Within any given population, specimens without markings always occur. These usually have a back the colour of verdigris and bronze-coloured sides. In these specimens the reticulated pattern is barely recognisable. The underside, which does not have spots, is white, but in some males may be

Podarcis peloponnesiaca

orange or brick red. The four subspecies are distinguished by differences in colour, markings, and scalation.

Habitat: Dry areas from the seashore up to an altitude of around 1,600 m.

Lifestyle, husbandry, and reproduction: This species does not usually climb and is mainly a ground dweller. In the wild its food consists mainly of insects, spiders, and woodlice. In the vivarium they will greedily eat crickets, grasshoppers, mealworms, flies, maggots, ants and their pupae as well as all manner of other insects and also earthworms.

Males are very aggressive towards one another. In spring and summer their period of activity begins shortly after dawn. As soon as the rays of the sun reach their refuges, they emerge to bask in their favourite places. When the weather becomes too hot the lizards retreat into their favoured hiding places. Mating has been observed as early as the beginning of May. It would appear that there are montane populations in which the females lay only one clutch each year, whilst on the lowlands two clutches are more usual. Each clutch invariably contains three or four eggs with average dimensions of 16.4×8.7 mm. At a temperature of 28 °C, incubation takes around 40 days. Two young that were bred in captivity had SVLs of 63 and 67 mm. In appearance, juveniles are very similar to adults, but their seven longitudinal stripes are brighter and more prominent.

During the summer months, this species is best kept in an outdoor enclosure before being returned to their indoor vivarium in early autumn.

Podarcis perspicillata (Duméril & Bibron, 1839)
Spectacled Wall Lizard

Distribution: Northwest Africa (Morocco, Algeria). In Europe only on Menorca where it was introduced.
Description: This species reaches a SVL of only around 5 cm, with the tail being twice as long. It differs from other European Wall Lizards in that it has developed a transparent "window" on the lower eyelid.

The compressed body is olive to bronze, green to blue-green, or blackish brown with two broad, light, longitudinal stripes as well as both light and dark flecks or a reticulated pattern. Specimens without markings are also occasionally found. The white, greenish, or blueish underside often has black spots.
Habitat: Mountain slopes, cliffs, and scree-covered hillsides. Amongst human habitation. These lizards are never found in very dry areas.
Lifestyle, husbandry, and reproduction: This species is a very adept climber on the branches of trees. Their food consists mainly of insects, especially flying insects which they catch expertly. They will however also eat ants and fruit during the dry season. Fruit satisfies their liquid requirements. Mating takes place in spring with the eggs being laid in June. The eggs measure 11.5 to 15.0 × 7 to 8 mm. The young hatch in July or August. It would appear that hibernation is not needed since in the wild this species has been observed throughout the year. A female owned by Richter (1986b) laid two clutches, each containing three cylindrical eggs measuring 15 × 8 mm. Unfortunately these eggs were not successfully incubated.

Podarcis pityusensis (Bosca, 1883)
Ibiza Wall Lizard

Distribution: The Spanish island group: The Pityusas, Ibiza, Formentera, and some nearby islets. Also found on the small island of Los Isoletas near the Balearic Island of Majorca.
Description: These lizards only rarely reach a SVL of 7 to 8 cm; the tail is twice as long. They have relatively rough, six-sided, and lightly keeled elongated dorsal scales that form fewer than 70 rows. This species differs from Balearic lizards in that the latter have 70 to 90 rows of scales. The colouring of these lizards is so extremely variable that 30 subspecies have been described. The back is usually green, brown, or grey, although males tend to be a greenish colour whilst females tend more to brown. The markings frequently consist of three brown to black stripes. There are rows of dark flecks on the flanks. The outer rows of ventral scales have a blueish tone. The underside may be grey, yellow, green, or red. The throat is usually flecked. The

Ibiza Wall Lizard also tends to melanism, so that specimens with a black, dark brown, or blueish back and a blue underside are frequently found.
Habitat: Dry stone walls, bushes and hedges, waste land, and fallow fields.
Lifestyle, husbandry, and reproduction: This species not only eats all manner of insects, but also a large amount of plant material. They should be kept in the same way as *Podarcis milisellensis*.

Frommer (1984) kept a pair of *Podarcis pityusensis* with an additional male in the vivarium. In May the lizards were placed in an outdoor vivarium measuring 70 × 35 × 50 cm which they shared with a male and two females of *Larceta agilis*. The furnishings were a tree root, pieces of bark, clumps of moss, and ivy. The substrate was a layer of forest loam. There was no aggression between the occupants of the vivarium. They were fed on field crickets, mealworms, grasshoppers, and spiders and were fed twice weekly. From mid-September the lizards were no longer visible and had probably gone into a form of "winter rest." They were, however, left outside until the end of October, after which they were placed in an indoor vivarium measuring 80 × 40 × 60 cm at a temperature of 25 to 30 °C. Beneath a spotlight the localised temperature was 35 °C. Several other species of lizards were also housed in this vivarium which was sprayed in the morning and again in the evening. On November 22, 1981, the *P. pityusensis* female laid three eggs in a flower pot containing garden loam. These were incubated during the day at temperatures between 28 and 32 °C and at room temperature during the night. On December 22, the female again laid three eggs. From January 18 onwards, i.e., after 57 days, the young began to hatch, and from January 22 they fed on small waxmoth larvae, small woodlice, and small crickets. The young from the second clutch hatched after 54 days, also within 3 days of one another.

Podarcis sicula (Rafinesque-Schmalz, 1810)
Ruin Lizard

Distribution: Apennine Peninsula and northern Italy, eastern Adriatic coast to Dubrovnik, Sicily, Sardinia, Corsica, Alba, and smaller Tyrrhenian and Adriatic islands as well as the Ile d'If off Marseilles and the coastal islands of the Sea of Marmara. Towards southeast Spain (Almeria), Menorca (Balearics). Introduced into the United States. Isolated populations live on the coast of North Africa in Libya and Tunisia. These lizards live on the coast as well as in the mountains up to an altitude of 1,300 m. On Mount Etna in Sicily, they live up to 1,800 m above sea level.
Description: These lizards are extremely variable in both colouring and markings. In the south of their distribution range they reach a SVL of up to 9 cm, whilst in the north the maximum is around 7 cm. At present over 40 subspecies are recognised. These subspecies show enormous differences in colour and markings. Along the back and flanks, Ruin Lizards may have all variations of green, olive, yellow, grey, and brown tones. Dark brown flecks and whorls run together to form longitu-

Podarcis sicula campestris

dinal stripes or a reticulated pattern. The longitudinal striping is usually more prominent in females. Animals from the north of the distribution range have especially vivid, light stripes along the back. There are, however, also black specimens and others without any markings. The underside, which is not normally flecked, is usually white to yellow, but may be green or various shades of red, even brick red.
Habitat: Grassland, roadsides and escarpments, in gardens, amongst rocks, on refuse dumps, on walls, and in similar places.
Lifestyle, husbandry, and reproduction: For many years, Ruin Lizards were sold in pet shops as "food lizards" for snakes and other carnivorous animals. They feed on insects and spiders. Some will also eat plant material. Ruin Lizards hibernate from October or November until February or March. Females lay eggs from April to June in holes that they dig themselves beneath bushes, tree roots, or in wall crevices. In especially favourable climates, up to five clutches may be laid in 1 year. A clutch usually consists of 3 to 12 eggs measuring 10 to 12 × 5 to 6 mm. Sexual differences develop in the egg after 11 to 13 days at 26 to 27 °C. The young hatch after 9 to 11 weeks between June and September. On hatching the young measure from 56 to 65 mm, and they are sexually mature at 2 years of age. It has however been known for these lizards to mate at an age of only 3 1/2 months. At this age the male and female had a SVL of 6.5 and 6.1 mm respectively (Henle & Klaver, 1986).

Reiss (1961) kept a female *P. sicula coerula* with a male *P. sicula sicula*. After hibernation and at temperatures between 8 and 14 °C the lizards mated in mid-February in complete darkness. During the evening of March 5, the female laid four eggs at the end of a tunnel that she had excavated. The tunnel was finally sealed by the female using head, snout, and feet. The female frequently visited the place where the eggs had been laid. During the following day the eggs were transfered to a flower pot containing garden loam and were returned to the vivarium. During the day the temperature varied between 23 and 24 °C, falling at night to 15 to 17 °C. Three young hatched after 82 days and were then placed in a rearing container where they fed on aphids, spiders, small crickets, and other small insects. They sloughed for the second time after 14 days and began to assume a green sheen that became more intense after each slough. One of the males was sexually mature at the age of only 10 months. The larger of the two females laid three eggs on March 14 of the following year although she was not yet 10 months old. A second clutch followed around 1 month later. The young from the first clutch of eggs bore a resemblance to both parents and had reached the size of the parents at the age of only 17 months.

In Wittig's case (1987), a female laid two clutches of three and five eggs from which five young hatched. At the age of only 5 months, they had already reached a length of 18.4 cm with the SVL being 6.1 cm. Specimens that escape from an outdoor enclosure are sometimes able to find places where they are able to survive and reproduce. There is, however, the danger that they will interbreed with endemic lizards, producing undesirable intergrades. A population that was released at Dreisam (near Freiburg) in 1913 was still thriving in 1924. Unfortunately it did not survive the severe winter of 1928/29 (Trautmann, 1924).

Podarcis taurica (Pallas, 1814)
Crimean Wall Lizard

Distribution: Balkan Peninsula, northern Hungary, the southwest of the former Soviet Union, some Ionian Islands. Isolated populations are also found along the coast of northwest Asia Minor.
Description: This species may reach a total length of 25 cm, of which 16 cm is tail. In comparison to other species, the head and body of this species appear somewhat compressed. They differ from other species in their range by their toothed collar. The forepart of the unflecked body is more-or-less grass green to olive green, whilst the flanks and hindquarters are brown with dark brown to black flecks. Along each side there are two white-to-yellow stripes. Flecks and stripes may be absent in animals from the south of the distribution range. There are often blue flecks on the flanks and shoulders. In females the underside is white without any markings, whilst that of males is orange, red, or yellow. Juveniles are much more vividly striped and do not have the green on the back.
Habitat: Grasslands that are not too dry, parks, roadsides, forest clearings, and sand dunes with vegetation. Also found on cultivated land and areas overgrown by bracken.
Lifestyle, husbandry, and reproduction: These lizards rarely climb and are never

found on cliffs. In May and June the females lay between two and six eggs measuring 9 to 15 × 6 to 8 mm. In the wild, the young hatch after around 2 months and reach sexual maturity after the second hibernation. They should be kept in the same way as *Podarcis melisellensis*. In captivity these lizards quickly become quite tame. If they are placed in an outdoor enclosure however, they quickly lose their tameness. In the wild, orthoptera, especially small grasshoppers and crickets, form the major part of their diet.

The Crimean Wall Lizard is one of the most highly recommended species and may be kept in a south-facing outdoor enclosure containing shrubs, grasses, rock piles, and climbing branches. Insects that enter the enclosure will not provide sufficient food for the lizards however, so they must be given additional crickets, grasshoppers, and other insects. The enclosure should be sprayed daily. In Magdeburg (central Germany), animals that had been released not only survived the winter, but also produced three young during the following year. The garden into which they were released contained shrubs and a rock garden (Petzold & Wehner, 1965).

Podarcis tiliguerta (Gmelin, 1789)
Tyrrhenian Wall Lizard

Distribution: Corsica, Sardinia, and neighbouring islands.
Description: This attractive species reaches a SVL of 6.5 cm. The ground colour on the back is brown, especially in females, or partially green. These lizards have prominent light stripes along the sides of the back. The black patterning may form stripes or a reticulated pattern. It may, however, be completely absent. Some island populations are very dark in colour. Males frequently have blue flecks on the flanks and shoulders. The underside is white, yellow, orange, or red with dark flecks, especially on the throat.
Habitat: Dry areas with dense vegetation. Scree-covered hillsides and dry stone walls. On cliffs and the edges of fields up to an altitude of around 1,800 m.
Lifestyle, husbandry, and reproduction: See *Podarcis melisellensis*. In the wild, hibernation lasts from late autumn until late February or early March. This species mates in May when males vigourously defend their territory. Females lay between 6 and 12 oval eggs measuring around 16 × 10 mm. In the wild the young hatch after 2 to 3 months.

Podarcis wagleriana (Gistel, 1868)
Sicilian Wall Lizard

Distribution: Limited to Sicily and the Aegean and Liparic Islands. This species is absent from northeast Sicily.
Description: These lizards usually have a SVL of 7.5 cm with the tail being twice as long. The upper side of males is usually green whilst that of females is more olive or brown. There may be one to three longitudinal rows of dark spots. The sides have

Lizards

Podarcis tiliguerta

Podarcis wagleriana

dark spots or are reticulated and frequently have blue flecks. Specimens without markings are also quite common. In females the underside is white, whilst males have a reddish underside with black flecks. There are always black flecks on the throat.

Habitat: Mainly on flat grassland. These lizards rarely climb on clifs, but will live quite happily on dry volcanic lava land. The limit of their distribution range lies at an altitude of 1,500 m.

Lifestyle, husbandry, and reproduction: In the wild, hibernation lasts from late October until March. Their diet consists of insects and spiders. From April to June the females lay four to six eggs measuring from 11 to 13 × 8 to 9 mm in holes that they dig themselves. The young hatch after 8 to 10 weeks measuring from 5.5 to 6 cm in length. They should be kept in the same way as *P. melisellensis*.

Psammodromus (Fitzinger, 1826)
Psammodromus, Sandrunner

The representatives of this genus live in southwest Europe and North Africa. *Psammodromus* live in open, sandy areas with sparse vegetation. The tall is double the SVL. Dorsal scales are large and usually heavily keeled. The collar is either absent

Psammodromus algirus

or reduced to a fold of skin in front of the forelegs. These lizards can emit a squeaking sound.

Psammodromus algirus (Linnaeus, 1758)
Algerian Psammodromus, Algerian Sandrunner.

Distribution: Iberian Peninsula and northwest Africa.
Description: This species reaches a total length of 27 cm. Along the sides of the bronze-coloured body there are usually two ochre or whitish yellow stripes bordered by two dark brown or blackish brown bands. There is often a dark central stripe along the back. The underside is white or light green. Males often have individual blue scales and large blue spots on the forelegs. During the breeding season the throat and sides of the head are orange to red. The orange tail of juveniles is particularly noticeable.
Habitat: On and amongst bushes such as citrus, gorse, and bramble and in areas devoid of trees.
Lifestyle, husbandry, and reproduction: These lizards feed mainly on insects and spiders which they catch amongst the undergrowth. They climb expertly amongst low bushes. Mating begins in early spring after hibernation. Females lay 8 to 11 eggs in late June or early July. A second clutch may be laid in late summer.

Because of their active lifestyle, *Psammodromus* require a spacious vivarium with a 5 to 10 cm layer of sand as substrate. Climbing branches, rocks, cork bark, and dry leaves should also be provided (Laube, 1981, In den Bosch, 1986). During the day the temperature should be 26 to 30 °C, localised beneath a spotlight to 35 to 38 °C. At night a temperature of 20 to 22 °C is adequate. *Psammodromus* feed on the usual mixture of insects, showing a distinct preference for soft-bodied arthropods. A hibernation of 2 to 4 months is important, not only to stimulate breeding condition but also for the general well-being of the animals.

When In den Bosch moved his *Psammodromus* from an outdoor enclosure into an indoor vivarium, he was able to watch them mating immediately. Obviously the change of climate acted as a stimulant. The eight eggs of the first clutch were laid beneath a piece of tree bark; the nine of the second clutch in a 5 cm deep hole that the female dug herself in the moist substrate. The eggs measured from 13.8 to 16.6 × 9.3 to 10.3 mm. They were incubated at temperatures between 27 and 30 °C and high humidity. The young should be reared apart from the parents.

Psammodromus hispanicus (Fitzinger, 1826)
Spanish Psammodromus

Distribution: Iberian Peninsula, French Mediterranean coast, and lower Rhône Valley.

Description: This *Psammodromus* is significantly smaller than the Algerian variety and only reaches a SVL of around 5 cm. It is only on the sides of the neck that the remnants of a collar are still visible. The ground colour varies from brown to ash grey. Along the back there are usually white stripes or streaks or a black barred pattern. Specimens without markings are often found. During the breeding season, males have green flanks.

Habitat: Sandy areas with low bushes, shrubs, or grasses.

Lifestyle, husbandry, and reproduction: In the wild, this *Psammodromus* will often run very quickly over long distances. They quickly bury themselves into loose sand or beneath the roots of plants. They feed on all manner of insects, spiders, and small snails.

According to In den Bosch (1986), a dry vivarium measuring 60 × 30 × 30 cm is large enough to house a pair of these lizards. All other furnishings should be the same as those described for the Algerian *Psammodromus*. In the wild, these lizards hold and defend a territory of some 3 to 4 m^2. Breeding condition will be stimulated by a 2- to 3-month hibernation from December until late February when the animals should be kept at 5 to 6 °C and from 70 to 80% atmospheric humidity. Mating begins soon after hibernation. During the actual 40 seconds of copulation the male holds the female firmly in position by a bite to the back. A clutch consists of one to four eggs measuring 9.5 to 11.8 × 5 to 7 mm, and weighing 0.14 to 0.28 g. During the year, females may lay between three and seven clutches of eggs. At 29 °C the young hatch after around 38 days, whilst at 25 °C incubation takes around 56 days. One day before hatching, the egg begins to "sweat." After slitting the shell the young remain in the egg for several hours. On hatching they have a SVL of 19 to 23 mm with the tail measuring from 30 to 42 mm. When lizards of different sizes are housed together, cannibalism may occur. The young are darker and more brightly coloured than their parents.

Takydromus Daudin, 1802
Japanese Racerunners

These lizards are at home in the Amur region, in Japan, and over the whole of eastern Asia down to Indonesia. The genera *Platyplacopus* and *Apeltonotus* from the Riukiu Peninsula are considered by many herpetologists to be subspecies of the genus *Takydromus*. Some Racerunners live in moist, forested areas and fields. The varieties with longer tails live on open grassland.

These are very slender lizards; according to species the tail may be from two to five times as long as the SVL. The large dorsal scales are heavily keeled and form four continuous longitudinal rows. The scales along the flanks are small and granular. Only the side rows of ventral scales are keeled. There are lamellae on the toes. The collar is only very poorly formed or totally absent.

Takydromus sexlineatus Daudin, 1802
Six-Lined Racerunner

Distribution: Southern China, Hainan Island, Thailand, Laos, Kampuchea, Vietnam, northern Burma, and Malaysia.
Description: This lizard may reach a length of 36 cm of which 6 cm is SVL. On the upper side they are brown to green. Along each side of the back there is a yellowish white stripe that is bordered above by a narrow black stripe and below by a broad black stripe. The long toes end in short, hooklike, bent claws. The tail breaks off easily. Between the forelegs and hindlegs, males have around 12 irregularly arranged, black-bordered yellow spots. These are absent from females. Males also have a much thicker tail root.
Habitat: Grassland.
Lifestyle, husbandry, and reproduction: The long tail is often used to distribute the body weight over the largest possible area, enabling the lizard to run quickly over the grassland on which it lives. When moving quickly through grass, they will often lie the legs alongside the body and propel themselves by serpentine movements of the body and tail. Using the tail, they may also anchor themselves in the grass.

Hauschild (1986) kept a pair in a glass vivarium of $60 \times 40 \times 40$ cm. The substrate was a 20 cm layer of earth and leaf mould on a 1:1 basis. Indoor Bamboo (*Bambusa*

Takydromus sexlineatus (above), Takydromus smaragdinus (below)

sp.), green lillies, asparagus fern, and vine branches allowed the lizards to climb and provided hiding places. During the day the temperature was around 28 °C, reduced at night to 21 °C. The vivarium was sprayed daily giving 60% daytime humidity and 80% humidity at night. The lizards were fed on spiders, field crickets, flies, midges, and occasional meadow sweepings. The lights were switched on for 12 hours daily.

On August 14, Hauschild saw the first clutch of eggs being laid. The female dug a hole some 7 cm deep at the base of a branch. She then climbed the branch until her cloaca was around 2 cm above the hole. The two eggs produced were then covered with earth. The oval white eggs measuring 10 × 5 mm were left in the vivarium. At the point where the eggs were laid the temperature was 24 °C during the day and 20 °C at night. To prevent the eggs being eaten, crickets were no longer given to the adults. The first egg hatched after 46 days. The hatchling had a SVL of 15 mm and a total length of 60 mm. The back was olive-brown and there were pale stripes along the flanks. The underside was yellowish white and the tail bright red. The second young hatched 2 days later. At higher temperatures the incubation time is shorter. To prevent them being attacked or eaten, the young should be reared apart from the adults. Three days after hatching the young began to feed on Drosophila. Later they were given the usual larger food insects.

Takydromus smaragdinus Boulenger, 1887
Emerald Racerunner

Distribution: The islands of Okinawa, Miya Koshima, Amami, Oshima, and Kikaiga in the Riukiu Archipelago of Japan.
Description: These lizards reach a total length of 21 to 24 cm, with a SVL of around 5 cm. Whilst males have brown longitudinal stripes along the flanks and brown limbs, the entire upper side of females is a uniform green. The belly is whitish green. There is often a narrow white stripe from the tip of the snout along each side of the body ending at the hindlegs. The tail and centre of the back are covered by large heavily keeled scales, whilst those on the flanks are very small and even. Males may be easily recognised by the thicker tail root.
Habitat: Grassland.
Lifestyle, husbandry, and reproduction: Rössler (1992) kept several males and females in a 50 × 40 × 100 cm (L×W×H) vivarium. One-third of the floor area was water. The rear wall was covered in natural cork bark. Densely planted climbing branches completed the furnishings. Heat and light were provided by fluorescent tubes, each of 20-W, and a 20-W spotlight. These lizards are very voracious feeders and will eat all the usual insects.

During autumn and winter activity is somewhat curtailed, although these lizards are active and willing to breed throughout the year. Before mating, the male holds his partner firmly in position by a bite to the neck. During the actual mating the male uses the tail as a sort of "safety line." Copulation may last up to 10 minutes.

Females lay one or two eggs at intervals of 2 to 4 weeks. The average size of the eggs is 11 × 6 mm. Eggs are usually laid amongst piles of moss at the base of plants or climbing branches.

Despite relatively constant conditions, the incubation of the eggs laid by Rössler's lizards took between 32 and 51 days to hatch. During this time they absorbed moisture and increased in size to around 17 × 11 mm. On hatching the young had a total length of 6.5 to 8.0 of which 2.2 to 2.5 was SVL. Their colouring was similar to that of females. They had brown longitudinal stripes on the flanks and a pale orange tail. Although adults will not attack the young they should be reared separately and fed on small field crickets, house crickets, and other insects.

Tropidosaura Fitzinger, 1826
Southern Mountain Lizard

These small lizards have a short head, cylindrical body, and long cylindrical tail. There is no collar although some specimens may have a fold of skin on the neck. The body is covered in rough, spiny, overlapped scales. Femoral pores are present. The body is an unobtrusive brownish olive. Males develop much brighter colouring on the flanks during the mating season. These lizards live only on land, usually amongst rocks upon which they bask. They seek refuge in short tunnels that they dig themselves beneath rocks or in earth-filled rock fissures. This is also where they hibernate. A clutch may contain from two to seven eggs.

Tropidosaura gularis Hewitt, 1927
Flap-Necked Mountain Lizard

Distribution: The Cape; mountain folds of Matroosberg to Lady's Slipper near Port Elizabeth.
Description: Large lizards with a very prominent fold of skin on the throat. The subocular scales touch the lip. The scales on the sides of the neck are small, smooth, and granular, whilst those on the sides of the body are large, pointed, and overlapped. There are two large plates in front of the anus. The dark brown to olive-brown dorsal surface usually has a pale olive-brown stripe along its centre. The scales have black edges, giving the lizard a speckled appearance. There are also two whitish yellow side stripes which may be either poorly defined or broken to form flecks. Juveniles and females have a blue-grey tail; in adult males it is pale olive with black flecks. Males in breeding condition have a black head with yellow flecks; the yellow flecks on the flanks change to vivid orange and the dorsal stripe to green. This species reaches a maximum length of 20 cm but is usually somewhat smaller.
Habitat: Sandy areas with sparse heathland vegetation and overgrown mountain tops.
Lifestyle, husbandry, and reproduction: These lizards are quite common amongst the heathland vegetation on mountain tops. They are however, extremely timid and

difficult to catch. They are expert climbers that feed mainly on flies and beetles. In November, females lay between four and eight eggs in holes that they dig themselves beneath flat rocks. They should be kept in the same way as Racerunners (*Psammodromus hispanicus*) but should be given a 4- to 6-week winter rest period at temperatures between 15 and 20 °C.

Tropidosaura montana (Gray, 1831)
Smooth-Necked Mountain Lizard

Distribution: Cape Fold Mountains eastwards to the Cape and Amatola Mountains, Natal Midlands, and Drakensberg Mountains.
Description: These lizards may reach a maximum length of 18 cm but are usually around 12 to 15 cm in length. The head is short with only a poorly formed fold of skin on the throat. The tail is very long. The subocular scales touch the lips. The scales on the throat are keeled. The dorsal colour is olive-brown with a dark stripe down the centre and a greenish white to yellow stripe along each side. A broken white stripe runs from the lower lip along the flanks. Below this stripe there are large pale yellow flecks that change to orange in the breeding season. The underside is greenish white, often with dark flecking. During the breeding season the tail is light blue-green with black spots.
Habitat: Heathland areas and montane grassland.
Lifestyle, husbandry, and reproduction: These timid and secretive lizards often bask on plants on which they are superbly camouflaged. They are active during early morning and late afternoon, catching their food of small insects. In the wild, females lay four or five eggs in November. At a temperature of 28 to 30 °C, incubation takes 33 or 34 days. On hatching, the young are 56 to 61 mm long. These lizards should be kept in the same way as *Psammodromus hispanicus*. A winter rest period at temperatures of 15 to 20 °C is advisable.

SKINKS, SMOOTH LIZARDS
Family Scincidae

Skinks are mainly distributed throughout Southeast Asia, Australia, and Africa. Only very few species are found in Europe. The predominantly slender, elongated, usually cylindrically shaped skinks have a long tail and in most cases a pointed head that is barely set-off from the body. In all species the tail breaks off very easily. The tongue is broad and fleshy at the tip and is freely moveable. In all species the external ear openings are either considerably reduced in size or covered by scales. In most skinks the limbs tend to be reduced or even only rudimentary. Some species have normal five-toed feet, whilst in other species limbs are completely absent.

Giant Skinks
Subfamily Tiliquinae

Corucia Gray, 1855
Prehensile Tailed Skink, Solomon Islands Skink

As the single member of this genus, the Prehensile Tailed Skink (*Corucia zebrata*) lives mainly in tropical forests or as a follower of civilisation. It is also the largest of all living skinks. These powerful animals have a short, broad head with a relatively blunt snout and exceptionally large smooth scales, especially on the prehensile tail. The well-developed limbs have powerful claws.

Corucia zebrata Gray, 1855
Prehensile Tailed Skink, Solomon Islands Skink

Distribution: Solomon Islands of Bouganville, Choiseul, New Georgia, Isabel, Guadalcanal, Nggela, Malaita, San Cristobal, Ugi, Santa Ana, and Shortlands.
Description: These skinks reach a total length of around 75 cm and a SVL of 35 cm. They are pale olive-green with dark and light spots or stripes ("zebrata") along the back. The prehensile tail is brown or olive green.
Habitat: On Guadalcanal they occur from the northern, secondary coastal forests up to an altitude of over 1,000 m.
Lifestyle, husbandry, and reproduction: These skinks lead a very secretive lifestyle in trees, especially amongst the branches of fig trees (*Ficus* sp.). The common name refers to the prehensile tail which is used as an additional gripping organ. As a typical tree dweller they are able to take food from a bowl hanging from a branch by only the tail. The tail muscles allow the skink to climb back to the branch. When threatened these lizards hiss loudly, although they are normally quite peaceful animals. When in danger however, they will not hesitate to attack and can inflict an extremely painful bite (Peters, 1985). During the day they sleep individually on branches or in groups in hollow trees. They go to the ground when hunting for food at night. Their food consists mainly of the leaves of climbing plants (*Epipremnum pinatum*) and epyphitic pepper growths (*Piper* sp.). In captivity this species will eat lettuce, apples, pears, cherries, etc., the occasional boiled egg, and raw and cooked ground beef. They are especially fond of the leaves of *Monstera* and *Scindapsus* plants.

Because of their size, the Prehensile Skink requires a very large vivarium with many climbing branches. Croton plants (*Codiaeum* sp.), which the lizards will not eat, may be used for decoration. Each morning the vivarium should be sprayed with tepid water. These lizards will occasionally drink the resulting drops of water. A bowl of fresh drinking water must always be available. This will indeed be used for drinking, especially at night. During the day temperatures of 26 to 30 °C are necessary. At night, 16 to 20 °C is quite adequate.

Corucia zebrata

Honneger (1975) wrote about newly imported Prehensile Skinks. Soon after arrival, four of them died from a *Salmonella* infection and amoebic dysentery. The remaining animals were treated with Septicol™ and Flagyl™. A juvenile that was born 6 weeks after the arrival of the skinks also died as a result of *Salmonella* and amoebic dysentery. It had a SVL of 115 mm, a tail length of 182 mm, and weighed 125 g.

For 6 years, Alfred A. Schmidt (1991) kept a group of three of these animals. From 1986 to 1990 his seven females produced 22 young, comprising six sets of twins and ten individual young. At an average temperature of 25 °C the animals eagerly sought out positions beneath the high-pressure mercury lamp which produced a localised temperature between 28 and 34 °C. Daily spraying of the deep layer of moss on the vivarium floor produced humidity of 60 to 80%. Schmidt was also able to watch the animals mating. Females are gravid for 6 months and give birth to one or two young at intervals of 10 to 12 months. At birth the young weigh between 100 and 120 g and have a SVL of 14 to 16 cm. The total length is 29 to 33 cm. They may remain with the parents who protect and defend them against any aggressor.

Egernia Gray, 1838
Spiny Skink

Spiny Skinks live in Australia, usually on dry, stony ground, and grow to a length of between 45 and 55 cm. There are smooth-scaled species such as *E. inornata* and *E. major* and others with spiny, keeled scales. The former are reminiscent of Lacertids and the latter of representatives of the genus *Tiliqua*.

Egernia cunninghami

Egernia cunninghami (Gray, 1845)
Cunningham's Skink

Distribution: South and southwest Australia, southern Queensland.
Description: With a length of up to 40 cm, Cunningham's is one of the largest of all Australian skinks. Its build is somewhat reminiscent of the larger of the *Chalcides* species but their forelimbs and hindlimbs are much sturdier. In males, the colour ranges from an ochre yellow to a brownish black broken by fine yellow speckling. The dorsal surface of females is a salmon colour. The sexes may also be distinguished by the colour of the underside. In males it is a light reddish brown and in females more of a dark reddish brown.
Habitat: These skinks live in rocky areas, on hillsides, and at the edges of fields.
Lifestyle, husbandry, and reproduction: In their natural habitat, these diurnal skinks are usually found in groups. Individuals require a spacious vivarium with temperatures between 22 and 28 °C. A spotlight should provide a localised temperature of around 35 °C. Heating should be switched off at night. Firmly cemented rocks provide suitable refuges. Strong roots and branches must also be provided. A water bowl is essential. These skinks are sexually mature at around 2 to 3 years of age.

When acclimatised to the northern hemisphere they may be hibernated during autumn and winter, after which they will begin to breed. Before mating, the male bites the female firmly on the neck or above the foreleg, prior to inserting his tail beneath that of his partner. If the female remains lying with outspread forelegs and raised tail she is willing to mate. The actual copulation lasts between 15 seconds and

1 minute. Females that are unwilling to mate will run off with the male still firmly attached. Niekisch (1975, 1981) was able to watch several matings which always took place in March. The young were born around 100 days later. As no yolk sacs were found, Niekisch assumed that they had been eaten by the parents. Ebbert saw several times that the young of various *Egernia* species ate the placenta.

Hart (1978) also bred Cunningham's Skink in captivity. At the end of April a male and two females were placed in a simply furnished outdoor enclosure where they remained until mid-September. To protect the lizards from inclement weather, part of the enclosure was covered with rigid, transparent plastic sheeting. The animals were finally placed in an indoor vivarium containing rocks and cork bark. It was here that a female gave birth to 11-cm young which Hart reared in a container measuring 65 × 25 × 30 cm (L×W×H). They were fed on grated turnip, banana, strawberries, finely chopped beef heart, grasshoppers, and mealworms. Before being given to the animals the food was liberally dusted with powdered calcium. After only 5 months the young had attained a length of between 22 and 24 cm. Hart bred an additional three young during each of the following years. These were all produced by the same female. If she came too close, the other female was chased off by the male. The adult skinks were fed on grasshoppers, centipedes, beetles, strips of beef heart, a mixture of honey and yoghurt, raw egg, grated turnip, calcium, and dried Daphnia. They also ate nestling mice, fish, vegetables, fruit, clover, and dandelions (Zimmermann, 1983).

A female kept by Ebbert twice gave birth to five young in 1990. The young measured from 12.5 to 13 cm. Usually the young have a balanced sex ratio of 1:1.

Egernia depressa (Günther, 1845)
Pygmy Spiny-Tail Skink

Distribution: Southwest Australia
Description: The Pygmy Spiny-Tail Skink is fully grown at a length of around 15 cm. They differ from the previous species in their smaller size and shape of the tail which is considerably shorter and more compressed. There are three-spined scales along the back. According to their area of origin, the body is grey or orange with black and white spots or deep red with black spots.
Habitat: On bushy and grassy steppes, on the ground, or amongst the branches of low bushes.
Lifestyle, husbandry, and reproduction: To keep these animals properly, a roomy vivarium is needed. The simple furnishings consist of sand and gravel as substrate, several rocks, and some dry branches. These skinks are excellent climbers. The temperature should be between 22 and 28 °C, localised beneath a spotlight to 28 to 32 °C. These lizards require more heat than other species of this genus. They eat medium-sized insects as well as fruit and vegetables. Ebbert's animals underwent a summer rest period from early February until late March, during which time they

Egernia depressa

ate very little. In all other respects they should be kept in the same way as *Egernia striolata*.

Egernia hosmeri Kinghorn, 1955
Hosmer's Skink

Distribution: Northern Queensland and the border regions of the Northern Territories.
Description: This species reaches a total length of 30 cm and has a SVL of 18 cm. The dorsal surface is pale fawn to reddish brown. The head and neck are dark brown. There are pale brown or cream flecks and spots on the head, neck, and lips. On the back, flanks, limbs, and sometimes on the tail, there are scattered pale and dark brown spots. The underside is white, cream, or yellow. There are dark brown flecks on the throat.
Habitat: Dry, stony and rocky regions.
Lifestyle, husbandry, and reproduction: In the wild these skinks find refuge beneath rocks, in rock crevices and in the burrows of small mammals. They are excellent climbers and love warm sunshine. Because of the spiny scales on the back and flanks, these animals can clamp themselves into rock crevices and are rarely attacked by predators. They should be kept in the same way as *Egernia cunninghami*.

Egernia inornata

Ebbert kept 12 specimens, singly, in pairs, or as a breeding group, together with other Australian skinks in various-sized vivaria. The smaller vivaria measured 90 × 50 × 50 cm whilst the largest was 180 × 90 × 70 cm. The occupants of these vivaria were unable to see one another. Although these skinks are classed as diurnal and hunt for food early in the morning and in the early evening, they are still active for 1 or 2 hours after darkness. It is sufficient to feed these lizards at irregular intervals of 1 to 4 weeks. They will otherwise quickly become obese and lazy.

Egernia inornata (Rosen, 1905)
Desert Skink

Distribution: Arid and semi-arid regions in central Southwest Australia; eastern Yuna, Narembenn; in the Balladonia District and the south of the Great Victoria Desert. Also found in eastern Australia.
Description: With a total length of 16 to 18 cm, these smooth, stocky desert skinks are one of the smallest of all *Egernia* species. They have a powerful head which is well set-off from the body, short legs, and a tail that in cross-section is round and runs to a point. The tail exceeds the SVL by around 20%. These skinks are reddish brown with scattered light spots. The belly is white.
Habitat: Various desert and sandy regions.

Lifestyle, husbandry, and reproduction: These skinks usually live in small colonies in a series of tunnels that they dig at the bases of bushes, e.g., porcupine bushes (*Triodia* sp.). The tunnels have many branches and several entrances. These robust lizards are long-lived and very undemanding in captivity. There will be no aggression if they are given sufficient space and refuges. Ebbert keeps individual Desert Skinks in vivaria measuring 90 × 28 × 25 cm. The animals usually spend the daylight hours in their refuges and only emerge to bask. They frequently dig and will explore the entire vivarium daily. If they feel threatened, they will immediately flee into their refuges but reappear again quite quickly.

A suitable substrate is a 5 to 8 cm layer of a mixture of sand and fine gravel. The other furnishings are pieces of bark, dry grass and leaves, tree roots, and some rocks. A bowl of fresh water should be available at all times. During the day the temperature should be between 26 and 30 °C, localised beneath a spotlight to around 40 °C. At night the temperature should go down to around 18 to 20 °C. As well as insects, these skinks will eat fruit and vegetables.

Ebbert hibernates his Desert Skinks from early July until late September at 12 to 18 °C and in total darkness. After hibernation the lights are switched on for 2 hours on the first day and then gradually increased until they are on for 14 hours at the end of October. From mid-February to late March the animals have a summer rest period when they are only fed sparingly. Only gravid females or those that have recently given birth are given more food. The animals are only kept in pairs from late October until mid-November. The skinks usually mate immediately or fight. The male follows the female relentlessly around the vivarium until he can hold her in position with a bite. Females indicate their willingness to mate by holding the tail in an upright position. The actual copulation lasts around 1 minute and may be repeated after 1 1/2 hours.

Females are gravid for around three to four months after which they give birth to young measuring 73 to 82 mm. The length of gravidity depends upon the temperature in the vivarium. After birth the young eat the placenta. One day later they start eating the usual insects, pieces of apple and banana. When held against a strong light the young appear almost transparent. On each side of the tail, immediately below the cloaca, two longitudinal blood vessels some 2 to 3 cm long may be seen. These are continuations of the veins in the body. These blood vessels may be absent from some juveniles. Ebbert suspects that in this way, sexual differences may be determined at an early age.

Egernia major Gray, 1815
Land Mullet

Distribution: Australia: Queensland, New South Wales, Northern Territories.
Description: These skinks reach a maximum length of up to 75 cm but are usually somewhat smaller. Their glossy, dark brown body colour can appear almost black

Lizards

immediately after sloughing. Along each side of the body there is one light and one dark stripe. There are some dark flecks along the back and the belly is yellow to grey.
Habitat: These animals live near coastal rivers, beneath bushes, and between large rocks and boulders.
Lifestyle, husbandry, and reproduction: See *Egernia cunninghami*. These skinks can inflict a very painful bite. They hold on to the bitten person and twist the body several times along the lateral axis. They require a dry vivarium that is longer than it is high and that is very spacious to allow the skinks to move around freely. A suitable substrate is a mixture of sand and gravel. Firmly cemented rocks provide suitable refuges and hiding places. Dry branches should also be provided. The animals must not be able to burrow beneath the rocks which could collapse and cause fatalities. Females give birth to two or three young.

Egernia saxatilis Cogger, 1960
Brown Rock Skink, Black Rock Skink

Distribution: The highlands of southeast Australia, from western Victoria to southeast Victoria. Individual populations on the coast of northern New South Wales. The nominate form is designated "Brown Rock Skink;" the subspecies *E. saxatilis intermedia* as the "Black Rock Skink."
Description: These lizards have a SVL of 13 cm and a total length of 20 to 23 cm. Their black-brown upper side sometimes has a poorly defined pattern of dark stripes. The scales are keeled and the lips have whitish brown speckles. The belly is silver grey. The chest and belly are sometimes lemon yellow or orange. Head and body may occasionally have pale brown spots or flecks.

The black subspecies *E. saxatilis intermedia* has a blue throat with black spots. The scales on the upper and lower jaws are light and the underside is cream to pure orange.
Habitat: Rocky areas.
Lifestyle, husbandry, and reproduction: See *Egernia cunninghami*. *Egernia saxatilis*, which lives exclusively amongst rocks, finds adequate refuges in the rock fissures near which they hunt. They will also occasionally live beneath fallen trees. In the vivarium these animals require numerous climbing branches and tree roots. To feel comfortable they also require a great deal of heat. They should be fed once weekly, otherwise they will quickly become very thin and emaciated. This is a very intolerant species that is best kept in pairs or breeding groups. The young grow extremely quickly and attain sexual maturity at the age of 1 to 2 years.

Egernia stokesi (Gray, 1845)
Gidgee Skink

Distribution: Central Australia.
Description: The Gidgee Skink is fully grown at 22 to 28 cm. Like *E. depressa*,

they have a short tail that is compressed at the sides and has prominent spines. The body is less spiny than the tail. On the olive-brown body there are whitish flecks and wavelike transverse bands. Males frequently have a longer tail than females.
Habitat: Lives on open stony flatlands and on rocky slopes.
Lifestyle, husbandry, and reproduction: Essentially the same as *E. cunninghami*. The lizards kept by Zimmermann (1983) were fed on insects, snails, nestling mice, finely chopped beef heart, fruit, vegetables, and raw egg. The animals were kept in a dry vivarium at temperatures between 25 and 37 °C during the day and 18 to 22 °C at night. The highest temperatures were below the spotlight that had been fitted in the vivarium. Fine gravel and sand are both suitable substrates. Rocks, tree roots, and branches provide both decoration and refuges.

These sun loving skinks also exhibit imposing threat behaviour in the vivarium. If they feel threatened they will first give a warning by opening the mouth widely to show the dark blue tongue and red mucous membrane. If an attempt is made to handle them, they will not hesitate to bite, at the same time lashing the tail from side to side. Their senses of smell and taste are obviously well developed. As soon as food is presented they will scurry to the bowl to investigate its contents. When they reach sexual maturity at around 3 1/2 years of age, the young must be separated from the breeding group. This is especially true for female young which will be viciously attacked by older, more dominant females. Breeding animals of around the same age is quite easy. The breeding season lasts from early September until late October. The male follows the female around the vivarium licking her tail and cloacal region before biting her firmly on the neck. When the female is held flat on the ground and raises the tail, this is an indication that she is willing to mate. Copulation lasts up to 15 minutes. The female is gravid for around 3 1/2 months before giving birth to from one to five young which measure 8 to 8.5 cm and weigh around 6.0 g. Ebbert's young ate the placenta shortly after being born. After sloughing, they also ate the discarded skin. For safety reasons, the young should be reared apart from the parents. One of Ebbert's females laid some waxlike eggs at the end of January 1990. These she subsequently ate.

Egernia striolata (Peters, 1870)
Tree Skink

Distribution: Australia: eastern and central Queensland through New South Wales—except on the coast and at higher altitudes—to Victoria and eastern Southern Australia.
Description: The upper side of this 22 cm skink is usually a dark brown-black to grey-brown with a wide, pale stripe running from the head to the root of the tail. On the lower reaches of this stripe there may be some dark scales. The pale dorsal stripe may have some black scales along its entire length. In most specimens, the dorsal surface is also speckled with white spots and flecks. The scales on the head

Lizards

Egernia stokesi (above), Egernia striolata (below)

are dark-edged and occasionally have a small number of white spots. The lips are white to cream with dark brown edges. A broad dark band runs along the side from the eye to the tail base. This stripe may also contain a number of lighter coloured scales which may form a transverse barring pattern. The underside is silver grey with dark brown barring or flecks on the throat, chest, and flanks. The underside of the tail and inside surfaces of the limbs are white to grey with dark speckling. The underside may occasionally have a hint of yellow or orange.

Habitat: As tree dwellers, these skinks may also occasionally be found beneath loose bark, in holes and hollows, sometimes below fallen trees, or amongst piles of rocks.

Lifestyle, husbandry, and reproduction: For these skinks, a vivarium of 60 × 35 × 60 cm is quite adequate. The substrate should be a layer of fine gravel on which cork bark, branches, roots, succulents, and a drinking bowl are placed. Ebbert's animals hibernate in complete darkness from July to September at temperatures between 12 and 17 °C. The new season begins in October with the vivarium lights being switched on for 6 hours daily. By mid-November, this has increased to 12 hours daily whilst the temperatures have been gradually increased to 22 to 26 °C during the day and 16 to 20 °C at night. During this time the skinks should be well fed. By mid-January, the lighting has been increased to 14 or 15 hours daily and the temperature to over 30 °C. At this time the vivarium should only be sprayed every 2 weeks, and the animals fed very sparingly. From May to late June the lighting is once again slowly decreased until it is finally switched off. A suitable lighting source is a 30-W neon tube, which from October to January and May to June is supplemented by a 40-W spotlight. From January to May the power of the spotlight should be increased to 60-W (Ebbert).

These skinks eat pieces of apple and orange, kiwi fruit, strawberries, cherries, banana, paprika, dandelion, tomato, beetles, field crickets, grasshoppers, and nestling mice.

During October, Ebbert was able to see his animals mating on several occasions. The male follows the female around the vivarium biting her on the flanks and at the root of the tail. He then stealthily approaches the female, quickly vibrates the tip of the tail, then suddenly seizes her by the neck holding her firmly in position. Mating lasts from 1 to 3 minutes.

On January 16, 1989, two young were born between 1200 and 1300 h. They were were quite sturdily built and were 10 cm long. The third young was born the following day at 1040 h. This young was born head first. Immediately after being born, it broke the umbilical cord which it ate together with the placenta. The entire process lasted around 1 minute. On January 18, at 1000 h, a fourth young, 10.5 cm long, was born. On each occasion the female gave birth whilst on a branch. In no case did the offspring fall to the ground. At an age of only 1 year, the young had already reached a length of 16 cm.

Egernia whiti Lacepede, 1804
White's Skink

Distribution: Over the whole of southeast Australia from the southern Australian peninsula to the border of Queensland. Also found in northern Tasmania.
Description: These large skinks reach a length of 30 to 35 cm. They have a long tail which runs to a point, and smooth scales on the body. On the bronze-brown dorsal

surface they have two black longitudinal stripes broken by white flecks. Along the sides of the body the white flecks have black borders. The tail region is pinkish red. The scales on the ear openings and those above the eyes are white. The underside is light olive to blue green.

Habitat: Mainly in rocky coastal areas. Also hollow tree stumps, rock crevices, grassland, on heathland, and woodland.

Lifestyle, husbandry, and reproduction: This is a rather intolerant species which requires a dry vivarium that is taller than it is long. The vivarium may be decorated with firmly cemented rocks that will also provide refuges. Occasional spraying is advisable, after which the skinks will lick the wet rocks. A shallow bowl of drinking water should always be available. These diurnal skinks require temperatures between 25 and 28 °C, localised beneath a spotlight to 28 to 30 °C. At night the heating should be switched off, allowing the temperature to fall significantly. Temperature variations are essential to encourage breeding. If kept at constant temperatures, these skinks will not come into breeding condition (see *Egernia striolata*). Females give birth to three or four young. These skinks eat various arthropods, occasionally snails as well as fruit and vegetables, but may at times be very selective in their eating habits.

Tiliqua Gray, 1825
Blue-Tongued Skinks

Blue-Tongued Skinks live in the semi-deserts, bushy steppes, and light forests of Australia, Tasmania, and New Guinea. They are also found on some Indonesian Islands, where they live in cultivated fields and moist forests. According to species they may reach a length of 15 to 55 cm. Their cylindrical, stout body is compressed and has smooth scales. The cone-shaped head is relatively large. Characteristic features are the blue tongue and the red mucous membrane. *Tiliqua gerrardi* is the only species with a pink tongue, but on the basis of several other features it is still assigned to the same genus. The well-developed limbs are usually very short. Earlier known as the Pine Cone Skink it was assigned its own genus (*Trachydosaurus*).

Tiliqua casuarinae (Duméril & Bibron, 1839)
Oak Skink

Distribution: The eastern coastal regions of Australia.
Description: These skinks reach a total length of up to 35 cm. Their somewhat serpentlike body has extremely short legs and is pale grey, pale fawn, olive-brown, or reddish brown in colour. On the body and flanks, pale-coloured scales form longitudinal stripes that become even paler in the centre. Along the flanks there is usually a pattern of transverse bands. The underside is pale grey to olive green. Occasionally the throat has dark flecks. The chest, hindquarters, and tail have prominent

dark-edged scales that form a dark transverse pattern. In juveniles there is a a wide collar from ear to ear.
Habitat: In slightly moist meadows, in the vicinity of ponds and brackish lagoons, on coastal heathland and sand dunes to dry forest.
Lifestyle, husbandry, and reproduction: These skinks are usually active at dusk and during the hours of darkness. They are normally found beneath fallen leaves or tree trunks. Because they may be extremely aggressive towards one another it is best to keep them individually, placing them together—under supervision—only during the breeding season. For a single skink, a vivarium of 50 × 25 × 30 cm will suffice. A 5 cm layer of sand should be provided as substrate. This species should be kept in the same way as *Tiliqua gerrardi*. Mebs (1974) fed his skinks on small snails, raw meat, and fish which they took from forceps. Mealworms were also greedily eaten. Females give birth to up to six young.

Tiliqua gerrardi (Gray, 1845)
Pink-Tongued Skink

Distribution: Eastern Australia, north of Sydney in the area around Gosford, New South Wales to Cape York in Queensland.
Description: Pink-Tongued Skinks reach a length of up to 40 cm. The upper side of the body is silver grey to brown with numerous dark brown transverse bands. The underside is grey, yellow, or orange. The pink tongue is particularly conspicuous.
Habitat: Amongst low vegetation in semi-moist and dry woodlands.
Lifestyle, husbandry, and reproduction: Pink-Tongued Skinks require a semi-moist vivarium with daytime temperatures between 20 and 25 °C, falling at night to 14 to 18 °C. The animals enjoy daily spraying. A mixture of earth and peat around 10 cm deep is a suitable substrate. Bundles of moss, climbing branches, tree roots, and some plants complete the furnishings. A bowl of fresh water should be available at all times. At dusk and during the night, the skinks will wander around the vivarium searching for food. Using their strong claws and mobile tail, they are excellent climbers. By flicking the tongue in and out they are able to detect minute scent particles. As soon as they discover a snail it is quickly seized and taken to a safe place for consumption. To reach the soft body inside, the snail shell is repeatedly beaten against a solid rock or other object until it breaks. Pieces of shell are also often ingested. Although these lizards may be given insects, spiders, chicken liver, chicken heart, beef heart, and occasional pieces of banana, their main diet in the wild is snails and slugs with which they should also be provided regularly if they are to remain healthy in captivity. A supply of slugs and snails should be kept for the winter months. Suitable snails are *Cepaea nemoralis* and *Cepaea hortensis*, which may be kept in a deep freezer. If other insects are smeared with snail slime before being presented, they will also be greedily eaten by these skinks (Zimmermann, 1983).

Pink-Tongued Skinks of the same sex will often fight when kept together. In such

Tiliqua gerrardi

cases, several retreats must be provided. It is much better however to keep this species in pairs. During the breeding season the male follows the female relentlessly for several days, attempting to bite her, until she finally submits to his advances. The male holds the female firmly in position by a bite to the neck or lower foreleg. Around 5 months after mating, the female gives birth to up to 25 young which are between 6 and 10 cm long. The juvenile markings consist of black transverse bands on a light brown to reddish background. It is essential that the young are reared apart from the parents. Breeding is much easier if the animals are hibernated beforehand.

In April 1985, Hauschild (1988) obtained two 1-year-old males and one female. They were placed in small individual vivaria and in complete darkness, at around 18 °C from August 28, 1985 until October 15, 1985. They were then placed together in a vivarium of $80 \times 60 \times 60$ cm. They began feeding 1 week later. On April 20, 1986, the female gave birth to seven young, each measuring 6 cm in length. Pink-Tongued Skinks are sexually mature at around 1 1/2 years of age.

In summer the parents were placed in their winter quarters where they remained until November 1, 1985, when they were again placed together in the larger vivarium. Mating took place almost immediately, and after 5 1/2 months and a dose of Oxytocin™ the female gave birth to 16 young. It has proven expedient to feed the young individually.

Tiliqua multifasciata Sternfeld, 1919

Distribution: The northern half of Western Australia, through northern South Australia and the Northern Territories to western Queensland.
Description: These skinks attain a SVL of around 30 cm. The tail adds a further 40 to 50%. They have a pale grey to grey-brown dorsal surface with several broad, pale orange-brown transverse bands on the neck, body, and tail. The transverse bands are broader than the spaces between them. Towards the centre of the back they become somewhat narrower. There is a black stripe from ear to ear. The black hindlegs have poorly defined orange-brown transverse bands. These short-legged skinks have a white or cream underside.
Habitat: Dry to semi-dry areas, rocky hillsides, and sandy deserts.
Lifestyle, husbandry, and reproduction: These ground-dwelling skinks may be kept in the same way as *Tiliqua scincoides*. Females give birth to up to 10 young.

Tiliqua scincoides (Shaw, 1790)
Blue-Tongued Skink, Common Blue Tongue

Distribution: Northern and eastern Australia.
Description: These Blue-Tongued Skinks reach the considerable size of up to 50 cm. The grey or light yellow to reddish-brown body has dark brown to black transverse bands. There is also a stripe from the eye to the nostril. The underside is yellow.
Habitat: Semi-deserts and bushland. Also found on cultivated ground.
Lifestyle, husbandry, and reproduction: Because of their size, these skinks should be housed in the largest possible vivarium which should be kept dry. Sand is a suitable substrate. Stacked rocks that are firmly cemented together and some individual rocks will provide adequate refuges. During the day, temperatures between 25 and 28 °C are necessary, whilst at night 15 to 22 °C suffices. The substrate should be evenly heated by a heating cable or heating pad. These diurnal skinks are great lovers of heat and will often be found basking beneath a spotlight that provides a localised temperature of 35 °C. A bowl of fresh drinking water should always be available.

When they feel threatened, these skinks open the mouth widely to show the blue tongue. Whilst doing this, they will also hiss loudly. These fairly aggressive skinks are best kept in pairs. Before mating takes place, the male sometimes follows the female around the vivarium for several hours until he is able to hold her firmly in a copulation bite. During the 4 month gravidity, the female increases considerably in girth and may produce up to 25 young, each around 15 cm long. For safety reasons, the young should be reared in small groups in separate vivaria. They will quickly begin to eat field crickets, house crickets, and fruit. Apart from arthropods, adults will also eat snails, nestling mice, lean beef, fish, eggs, and canned cat or dog food.

Using a probe, Brauer (1980) was able to determine that his group of animals consisted of two males and three females. He placed one male and two females in a

Lizards   185

Tiliqua multifasciata (above), Tiliqua scincoides (below)

Tiliqua rugosa

140 × 45 × 45 cm vivarium. It was only after they had been hibernated, at 15 to 20 °C during the day and 5 to 10 °C at night, that courtship behaviour was seen. The result was four young 12 to 13 cm long and eight of 8 to 10 cm long.

By adding Sanostol™ to their drinking water, Schade (1977) was able to cure these skinks of a cold infection. He also bred these skinks successfully.

Tiliqua rugosa Gray, 1845
Pine Cone Skink, Shingle-Back

Distribution: South and southeast Australia.
Description: These skinks reach a length of 35 to 37 cm. They have greatly enlarged, keeled scales that protrude outwards from the body and only barely overlap. The light brown to black body colour sometimes has yellow flecks; the underside is yellow and black. Their cone-shaped scales, especially those on the short tail, are an unmistakable feature of this species. The tail serves as a fat reserve and thus has varying degrees of development from animal to animal.
Habitat: Desert areas.
Lifestyle, husbandry, and reproduction: In comparison to the body, the limbs remain too small for these skinks to escape from predators. To defend themselves, they bend the body into a semicircle, open the mouth widely, and emit a loud warning hiss whilst allowing the dark blue tongue to hang out. If they are approached too closely, they will not hesitate to bite and may inflict a very painful and bloody wound. Before mating the male follows the female relentlessly around the vivarium, licking her tail, cloacal region, and head before suddenly seizing her by the shoulder or foreleg. He then forces his body beneath that of his partner and copulation takes place. Within 1 year, a female produces two or three young, each around 14 cm long.

These young should be reared on small arthropods and sweet fruit. Adult Pine Cone Skinks eat all manner of insects, snails, nestling mice, raw eggs, fruits, and compotes.

This species requires a spacious vivarium and daytime temperatures between 22 and 30 °C, localised beneath a spotlight to 35 to 40 °C. At night the temperature should fall to 16 to 22 °C. Sand is a suitable substrate. Firmly cemented rocks will provide suitable refuges. Using Vybramycin™, Münsch was able to cure his Pine Cone Skinks of a cold infection. Vibramycin™ is a broad-spectrum antibiotic for children. Using a pipette, half the child dosage was administered over 4 to 5 days.

Although these skinks are easy to keep, they are not easy to breed if the animals do not have the climatic variations to which they are accustomed. In their natural habitat, the summer (from January to March) is hot and dry and food is very scarce. In autumn (from April to June), it is still warm and dry with little food to be found. During the cool winter (from July to September), there is no food at all. It is only in spring (from October to December), when it is again warm and dry, that sufficient food can be found.

It was under these conditions that Hitz (1983) kept two pairs of Pine Cone Skinks in vivaria measuring 100 × 60 × 40 cm. Twice weekly, for 20 minutes, the lizards were given ultraviolet light from an Osram Vitalux™ lamp. Around midday the humidity was around 43 to 46%. At other times it was higher because the vivarium was never sprayed! The "winter period" was spent in an outdoor enclosure measuring 100 × 40 cm. Here they received direct sunlight for only 4 hours daily, and were also subject to periods of adverse weather. They were fed twice weekly. One meal consisted of various fruits, and the other of various sorts of meat and crickets, both of which they ate greedily. Chopped chicken and banana were also given occasionally. One meal was supplemented with calcium phosphate, trace elements, and vitamins. During August and September no food was given, but drinking water was always available.

After their period outdoors the skinks were placed in the indoor vivaria at the beginning of October. A tube connecting the vivaria allowed the skinks access to both. One of the males became dominant and mated with both females 10 times between October 30 and November 9. From November 10, the females were no longer willing to mate. After being gravid for exactly 4 months, the first female gave birth to a single young on March 5 and again on March 15. The young sloughed within 2 hours after which they were placed together in a separate vivarium measuring 100 × 50 × 40 cm. This vivarium was furnished in the same way as that of the parents. Every 2 days the young were given the same food as the parents and twice weekly were given 10 minutes of ultraviolet light from an Osram Ultra-Vitalux™ lamp.

Gross (1989) also bred this species by subjecting the animals to climatic variations throughout the year. In summer he gave them a "winter rest" at 14 °C and on March 27 found a 17 cm juvenile in the vivarium although he had not had any indication that the female was gravid.

Skinks
Subfamily Scincinae

Chalcides Laurenti, 1768
Cylinder Skinks

Cylinder Skinks live in more-or-less dry areas of the Mediterranean region, in Africa to Somalia, and in Asia to India. Some species prefer moist meadows. The body, as the name would suggest, is cylindrical in shape and round in cross section. It is extremely elongated and slender. The head is cone-shaped. The tail is approximately the same as the SVL. The limbs are relatively short, with three or five toes, and are extremely small.

Chalcides bedriagai Bosca, 1880
Iberian Skink

Distribution: Iberian Peninsula.
Description: These skinks reach a total length of 16 cm. The upper side is light olive with the centre of the back being somewhat darker. There is a wide dark stripe along the flanks. There are irregularly spaced eyespots over the entire upper surface. The limbs have five fingers and toes.
Habitat: Very dry regions, more frequently near the coast than in the mountains.
Lifestyle, husbandry, and reproduction: These skinks lead a very secretive lifestyle and are active during the day and at dusk. In captivity these skinks gradually lose their timidity.

Because they can be rather aggressive outside the breeding season, the vivarium for a pair of these animals should have a ground area of at least 80 × 40 cm. One-half of the floor should be covered to a depth of 5 to 6 cm with sharp sand, whilst the other half contains a mixture of sand and earth in which plants may be placed. Stacked rocks and tree roots will be used for climbing as well as for providing hiding places. During the day the temperature should be 24 to 29 °C, localised beneath a spotlight to 35 °C. At night 20 °C is quite adequate. During the cooler part of the year these skinks should be hibernated at 6 to 10 °C. This will stimulate breeding. *C. bedriagai* eats spiders, beetles, snails, ants, woodlice, grasshoppers, and the eggs of insects. In captivity they will eat all manner of insects and spiders. The vivarium should be sprayed daily and a bowl of fresh drinking water should always be available.

This is a viviparous species. Several times each year females give birth to between one and four young. If housed in a separate vivarium, they are extremely easy to rear and should be kept under the same conditions as the adult animals.

Lizards

Chalcides chalcides Linnaeus, 1758
Three-Toed Skink

Distribution: Iberian Peninsula, southern France including Sicily, Sardinia and Elba. In North Africa between Morocco and Libya.
Description: The Three-Toed Skink reaches a total length of 48 cm, of which around 40% is tail. The greatly elongated, snakelike body has greatly reduced limbs, each of which has three toes. The brown, bronze, olive, or sand-coloured upper side has 9 to 11 dark longitudinal stripes. There are, however, specimens without markings.
Habitat: These skinks live amongst dense grass up to 40 cm high. Woodland or sparse vegetation form the borders of their natural habitat.
Lifestyle, husbandry, and reproduction: From spring onwards, these skinks are active throughout the day. During the hot summer months, they hunt for food during the early hours of the morning. When hunting for food they make use of their hindlegs, but when in danger they are able to move through the grass at amazing speed without using their legs. In the wild, these skinks feed on ground-dwelling insects. In captivity they should therefore be fed on meadow sweepings during the warmer months of the year. They will also eat the usual foods such as field and house crickets. The vivarium should be a little larger than for the previous species. In all other ways they should be kept in the same way as *Chalcides bedraigai*. Suitable furnishings are bunches of dry grass and tree roots. In spring and autumn the temperature should be around 22 to 28 °C, but in summer somewhat higher. Three-Toed Skinks should be hibernated from late September until the end of March. Mating commences soon after hibernation. Females give birth to between 3 and 15 young; on average however, seven is a more usual number. The young are more difficult to rear than other *Chalcides* species. It has proven more expedient to rear the young individually in vivaria with many hiding places. They should be given a large variety of small insects and should be fed daily.

Chalcides ocellatus Forskal, 1775
Ocellated Skink

Distribution: Sardinia, Sicily, Malta, and other small islands; in the Naples area, in Greece between the Peloponnes and the island of Euboea as well as on numerous small neighbouring islands. From Crete to Rhodes and Kios. In Cyprus. In North Africa and western Asia. From southeast Turkey to Pakistan.
Description: These skinks reach a total length of up to 30 cm. Adult specimens from Greece remain significantly smaller. The colour on the upper side ranges from light yellow through yellow-to-brown tones with a partially greenish sheen. The regularly arranged eyespots form transverse bands. There are hollows along each side of the body into which the skinks fold their legs when moving through sand in a serpentine manner.

Chalcides chalcides (left), Chalcides ocellatus (right)

Habitat: These skinks are much less demanding than the previous species. They live in the immediate vicinity of the coast amongst sand dunes. They are also found in the mountains, in cork forests, and at the edges of cultivated fields and olive groves. They are also frequently found near ruins and dry stone walls.

Lifestyle, husbandry, and reproduction: These skinks require much heat and are mainly active during the day. They are very timid and are fast-moving animals. In cooler climates they hibernate for up to 3 months. They are not selective in their feeding habits and will eat all manner of insects, smaller lizards, and even their own young. Females are gravid for 3 months after which they give birth to 2 to 25 young. The adults should be kept in pairs or in small breeding groups in the same way as described for *Chalcides bedriagai*.

In the vivarium these lizards will eat the usual insects as well as apple, banana, cherries, strawberries, fruit yoghurt (without preservatives), fine strips of beef heart, and ground meat (Harbig, 1986). They should be hibernated from the end of November until early February, during which time the temperature should be reduced to between 12 and 15 °C. When the temperature is again increased the male follows the female relentlessly. Heavily gravid females are best kept singly in individual vivaria, from which they should be immediately removed after the young have been born. According to the origin of their parents, at birth the young measure between 64 and 85 mm. They are extremely easy to rear. According to Houba (1957), females may also produce infertile eggs occasionally. In his case, the young from a female imported from North Africa measured between 64 and 67 mm.

Lizards

Chalcides polylepis Boulenger, 1897
Moroccan Skink

Distribution: Western Morocco north of Agadir.
Description: In appearance, this 30 cm skink is very similar to the Ocellated Skink (*Chalcides ocellatus*) making it sometimes very difficult to differentiate between the two species. In *C. polylepis* there is a very small ear orifice and 34 to 40 rows of smooth scales at the centre of the body.
Habitat: On flat and sloping ground in sunny, open spaces. In the Atlas Mountains they may be found up to an altitude of around 2,000 m.
Lifestyle, husbandry, and reproduction: In the wild these skinks may be found by turning rocks or discarded pieces of wood. At the slightest disturbance, these very timid creatures will retreat into their hiding places at an amazing speed. During spring and autumn these skinks are active during the day, whilst in summer they are usually only active at dusk and during the night. At the end of summer they retreat into their winter quarters where they remain until March. Mating takes place during April. Like *C. ocellatus*, *Chalcides polylepis* is ovoviviparous. In captivity, this species should be kept in the same way as *Chalcides bedriagai*.

Chalcides sexlineatus Steindachner, 1891
Striped Canary Island Skink, Gran Canaria Skink, Six-Lined Skink

Distribution: Gran Canaria, Canary Islands.
Description: This species only rarely reaches the maximum length of 18 cm. It is a slender cylindrical skink, the small head of which is barely set-off from the neck. The legs are very small and weak. Their colouring and markings are extremely variable. This is the only Canary Island skink with longitudinal stripes. The ground colour is light to blackish brown. The tail is either the same colour as the body or is a beautiful blue-green. Between the two variations, several transitional patterns are possible. The underside is yellow to orange, especially on the throat. Parts of the belly may also be blue-grey.
Habitat: Up to an altitude of 1,600 m. They are usually found on weather-beaten land with sparse vegetation. Here they take refuge beneath rocks.
Lifestyle, husbandry, and reproduction: A vivarium of 70 × 40 × 30 cm is large enough for a pair of these skinks. Additional requirements are the same as those described for *C. bedriagai*. During the cooler part of the year these skinks must withstand a temperature of only around 18 °C for some 5 weeks. They are rarely seen during this time.

When the temperature again rises, the male follows the female relentlessly, licking constantly until mating takes place. Whilst gravid the female steadily increases in girth and weight. After the daily spraying of the vivarium the female will eagerly lick up the resulting water droplets. Gravidity lasts from 3 to 4 months after which

Chalcides sexlineatus

the female gives birth to two or three young. These have a SVL of 38 to 40 mm and should be fed on meadow sweepings, small field crickets, and small house crickets. These skinks should not be overfed or they will quickly become obese. This condition is particularly noticeable from the shape of the tail. One feed per week is quite adequate for this species.

Chalcides viridanus (Gravenhorst, 1851)
Canary Island Skink

Distribution: Canary Islands. The nominate form lives on Hierro and Teneriffe as well as on the north coasts of Roque de Garachio and Roques de Anaga lying off Teneriffe. The subspecies *Chalcides viridanus coeruleopunctatus* lives on Gomera.
Description: These skinks may reach a total length of 18 cm. The top of the head, back, and often the tail are a copper to olive-brown colour. This is in great contrast to the dark brown to black on the sides of the head, flanks, legs, and tail, which in

most specimens is the same colour. In very few specimens with an olive-brown back the flanks may also be a lighter grey-brown. The centre of the body is covered with numerous small spots which may form up to 10 longitudinal stripes. The flanks and tail also have these small spots. The skinks of Hierro remain much smaller and are barely 14.4 cm long. They are however much daintier, darker, and more colourful although they do show a tendency to melanism.
Habitat: On Teneriffe up to an altitude of 2,000 m in areas of loose ground with sparse vegetation.
Lifestyle, husbandry, and reproduction: See *Chalcides sexlineatus*. These skinks are active during the day and at dusk. For the first few months in captivity they are extremely timid and secretive. They do however quickly become accustomed to their new surroundings. *Chalcides viridanus* should also only be kept in pairs.

Eumeces Wiegmann, 1834
Large Skinks

The skinks of the genus *Eumeces* live in the warmer areas of Asia, North Africa, and America. In appearance they are somewhat similar to Lacertids. They are however much smoother and have shorter limbs. Only a small number of American species give birth to live young. Most species lay eggs.

Eumeces fasciatus (Linnaeus, 1758)
Five-Lined Skink

Distribution: Massachusetts to Florida, westwards to Arizona, the Mississippi Valley to southern Canada.
Description: The Five-Lined Skink may reach a length of 20 to 25 cm. The tail is double the SVL. The delicate limbs end in very sharp claws. The young are especially striking because of their colouring and markings. On the upper side of the body there are three yellow longitudinal stripes. There is an additional stripe along each side of the body. With increasing age, these stripes become less noticeable until they eventually disappear. The blue tail also gradually becomes paler, whilst the head of males becomes a more intense red with increasing age.
Habitat: Mainly on the ground in light, hilly woodlands. These skinks will climb on rocks or fallen trees in search of insects.
Lifestyle, husbandry, and reproduction: The Five-Lined Skink requires a semi-moist vivarium that is longer than it is tall. Below a spotlight there should be a localised temperature of 35 °C whilst elsewhere temperatures between 23 and 30 °C are adequate. At night all heating should be switched off allowing the temperature to fall considerably.

A suitable substrate is a mixture of sand, peat, and earth. A flat rock placed beneath the spotlight will provide an ideal basking place. Pieces of tree branch, cork bark, and firmly stacked rocks will provide adequate and sufficient refuges. A large

Eumeces schneideri

form of "brood protection" by coiling their body around the eggs. They also turn the eggs regularly and bury them more deeply when it becomes too dry. If the immediate area around the eggs threatens to dry out, the female sprays the ground with urine. The young hatch measuring around 7 cm and are still protected by the female against possible predators.

Eumeces obsoletus should be kept in a spacious vivarium with a ground area of at least 1 m^2. The furnishings are tree roots and firmly stacked flat rocks that provide refuges. A suitable substrate is a 10 cm layer of a peat/sand mixture, one corner of which should be kept constantly moist. Great Plains Skinks eat all manner of insects as well as their eggs and larvae. In all other respects they should be kept in the same way as *Eumeces fasciatus*.

Eumeces schneideri (Daudin, 1802)
Spotted Skink

Distribution: From northwest Africa to western Asia.
Description: According to subspecies, these skinks reach lengths between 32 and 47 cm. They are very reminiscent of the genus *Chalcides*. Trapdoor-like scales close the ear orifices and protect the eardrum when the skinks are digging or burrowing in sand. Irregularly spaced orange-red and yellow flecks break up the otherwise brown background colour. Flecks are also found on the yellowish-coloured sides. Along the backs of these skinks there are orange-red dashes that alternate with black-bordered yellow scales. The underside is dirty-white to yellowish white.
Habitat: Mainly semi-desert and in dry, cultivated ares beneath rocks and bushes.

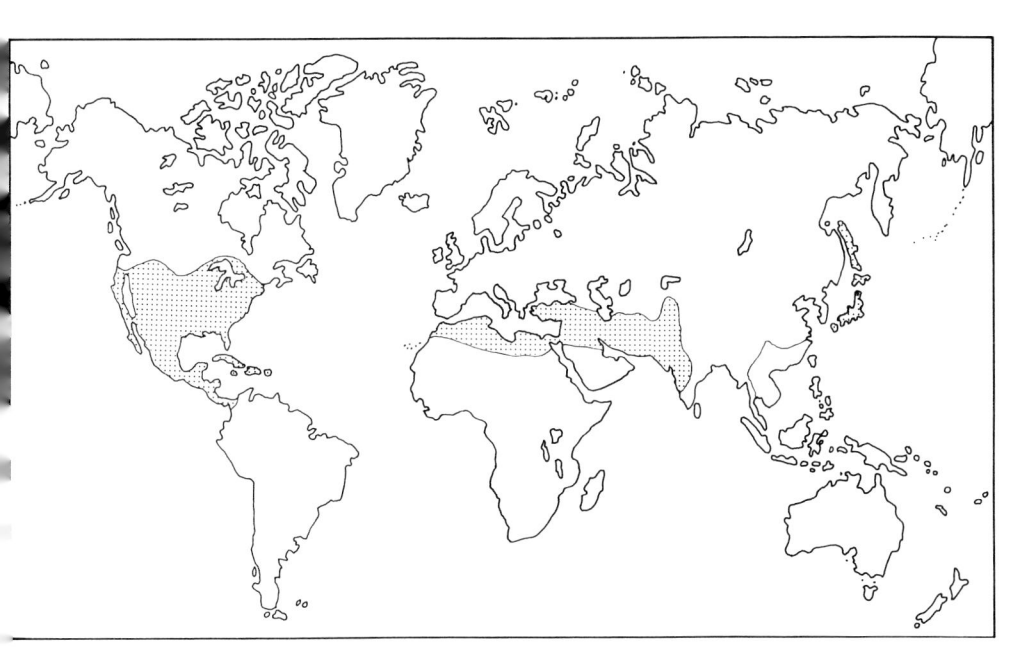

Distribution ranges of *Eumeces*

Lifestyle, husbandry, and reproduction: These diurnal skinks enjoy digging burrows in the ground. The burrow usually ends in a chamber in which they sleep. In captivity this is a very undemanding species. A pair requires a vivarium of at least 80 × 50 × 50 cm, the base of which should be covered by a 10 cm layer of a sand/earth mixture. A large, flat rock should be provided, for it is beneath this that these skinks will dig their burrow. Using a heating cable or a heating pad, the ground temperature should be kept around 28 to 30 °C. A spotlight should provide a localised temperature of around 35 °C. At night the temperature should be allowed to fall to approximately 20 °C (Thieme, 1983). The other furnishings required are pieces of tree branch, bogwood roots, and/or rocks. If plants are installed they must be left in pots so that the digging activities of the skinks will not damage the roots. In the vivarium, Spotted Skinks will eat all manner of insects, snails, nestling mice, ground beef, small freshwater fish, beef heart, liver, pieces of apple, and dandelion flowers (Kuchling, 1970). They will eat snails particularly greedily (Esterbauer, 1986). The vivarium must be sprayed every 2 or 3 days.

During the breeding season males are particularly aggressive towards one another. It is thus only possible to keep this species in pairs, or one male with two or three females. To determine whether she is ready to mate, the male follows the female around the vivarium licking her cloacal region. If the skinks are hibernated from October until early March it stimulates breeding. As soon as the temperatures slowly begin to increase and the skinks are given ultraviolet light for 10 minutes, twice weekly, the first matings will be seen. An Osram Vitalux™ lamp is an ideal ultraviolet light source. Around 5 to 6 weeks after mating, the female, which by this time has become extremely restless, begins to search for a suitable place to lay her eggs, normally somewhere moist, in a rock crevice or beneath a rock. The 3 to 20 parchment-like eggs are guarded by the female for the 8 or 9 weeks that it takes the young to hatch. The eggs should, however, be transferred to an incubator and should be kept at 27 to 29 °C.

Along the back, juveniles are grey-brown or olive-grey with 8 to 10 longitudinal rows of white spots amongst which there are scattered dark flecks. The orange coloured stripes are absent from the flanks. The underside, and often also the flanks, are a yellowish colour. The young should be reared individually on the greatest possible variety of food. This is a particularly easy species to rear.

Ophiomorus Duméril & Bibron, 1839
Snake Skinks

These worm-shaped lizards inhabit a distribution range stretching from Greece to western Asia. They have a wedge-shaped head with angular sides. The sides of the body are also angular. Some species have minute limbs that may be folded into hollows on the body. In other species the limbs are completely absent.

Ophiomorus punctatissimus (Bibron & Bory, 1833)
Speckled Snake Skink

Distribution: Limited to central and southern Greece, Peloponnes and the island of Kythia, and some small islands off Githion. Also found in Turkey in southwest Anatolia.
Description: These skinks may reach a total length of up to 20 cm, of which around half is tail. This is the only European skink that is completely without limbs. The yellow to brown colour on the back is often broken by small, dark brown spots arranged in longitudinal rows. These spots are much more noticeable on the tail and in the pelvic area than they are on the remainder of the body.
Habitat: Flat land in warm, dry valleys.
Lifestyle, husbandry, and reproduction: These Snake Skinks may only be found by turning rocks during the spring months. When disturbed they will retreat into their burrows at an amazing speed. The body moves in a serpentine manner. It is thought that they are only found at this time of year because frequent rainfall floods the burrows in which they live (Rödel, Bussmann & Kaupp, 1989).

A pair of Snake Skinks may be kept in a vivarium measuring 60 × 40 × 40 cm. A suitable substrate is a mixture of leaf mould and sand upon which a number of flat rocks and pieces of tree bark are placed. One part of the substrate should be kept permanently moist and a bowl of clean drinking water should always be available.

During the day, heat and light are provided by a spotlight. In this way the temperature varies from day to night between 35 and 18 °C. Because these skinks lead a very secretive lifestyle, the vivarium should be checked if the animals are not seen for a long period.

The Snake Skinks kept by In den Bosch (1988) were caught in the Peloponnes at the end of April 1984. During December and January, the vivarium was placed in a frost-free cellar at temperatures ranging from 0 to 8 °C. On June 1, three pinkish white eggs could be seen shining through the skin of one of the skinks. One of the eggs was laid on June 7 and the remaining two on June 8. After 43 days a dead young was found in the vivarium. It had hatched unnoticed from one of the two eggs that were laid together on June 8.

Scincopus Peters, 1864
Night Skinks

This group is closely related to the genus *Scincus* and contains as its sole representative the Night Skink *Scincopus fasciatus*. In appearance it is similar to and reminiscent of *Scincus scincus*, only the ear orifices are not so well covered. The head is large and wedge shaped. The tail is around half the SVL. The dorsal scales are heavily keeled with the two centre rows being greatly enlarged.

Scincopus fasciatus Peters, 1864
Night Skink

Distribution: Central and southern North Africa.
Description: The bright yellow ground colour becomes paler at the sides, whilst the underside is dirty-white. The almond-shaped eyes are surrounded by black scales. Along the back there are five to seven black transverse bands. The first begins at the neck and the last ends at the tail joint.
Habitat: Dry, sandy, scree-covered areas with sparse vegetation.
Lifestyle, husbandry, and reproduction: The activity period of this skink begins with the onset of darkness. During the day, temperatures between 25 and 28 °C are necessary. A spotlight should provide a localised temperature of around 35 °C. At night the temperature should fall to around 18 °C. This is achieved by switching off the heating in the early hours of the evening. The vivarium should be longer than it is tall and should have a substrate of smooth sand. Stacked rocks will provide refuges. A heating cable installed beneath the sand will provide various temperatures, enabling the skinks to seek out that which is more comfortable. The vivarium should be sprayed daily so that the skinks will satisfy their low liquid requirement by licking up the resulting water droplets. A short rest period at lower temperatures will not only stimulate breeding but will also contribute to the general well-being of the animals.

Johann (1980) kept three Night Skinks in a 120 × 60 × 50 cm wooden vivarium with large ventilation panels. They were fed thrice weekly on crickets, earthworms, grasshoppers, mealworms, and nestling mice. They only rarely ate pieces of melon or banana.

Scincus Laurenti, 1768
Sand Skinks

The distribution range of these smooth-scaled skinks stretches from northeast Africa to Iran. There, these powerfully built, cylindrical lizards live in deserts. Their cone-shaped head ends in an elongated, chisel-like snout. The ear orifice is almost completely covered. The tail is shorter than the SVL. The short, powerful, five-toed limbs end in broad, flat digits.

Scincus scincus (Linnaeus, 1758)
Sandfish

Distribution: North Africa. The northernmost limit of the distribution range is the south of the Atlas Mountains along the Mediterranean coasts of Tunisia, Libya, and Egypt through southern Israel, southern Jordan, and central Iraq into the southwest of Iran on the coast of the Persian Gulf. The southernmost distribution limit is from Senegal to Mali (Timbuktu), Adar des Iforas through northwest Nigeria, central Egypt (Dakhla), northern Sinai and southern Arabia to Yemen and the United Arab Emirates (Ras al Khaymah).
Description: The Sandfish is particularly well-adapted to its lifestyle. Its flat head and cone-shaped snout allow it to burrow into sand quite easily and quickly. Covers on the eyes and ears prevent the penetration of sand. In addition, in front of each nostril there is a buildup of sand that acts as a filter to prevent individual sand particles from entering the respiratory tubes. The broad, flat fingers and toes are especially suitable for burrowing. The short thick tail, which is flat on the sides, serves as a fat reserve. On a light yellowish-red ground colour these skinks have seven transverse bands along the back. Several subspecies are known.
Habitat: Dry, sandy areas, deserts and sand dunes in the Sahara Desert.
Lifestyle, husbandry, and reproduction: These diurnal skinks live exclusively on sand. They are usually found burrowed beneath the sand where they are heated by the rays of the sun. The feet, legs, and serpentine movements of the body and tail allow these skinks to burrow into the sand extremely quickly. Their food is not only animal, but also includes some vegetable matter such as seeds from winter rye (*Aritida pungens*).

Sandfish Skinks should be kept in pairs in a vivarium of at least 60 × 40 × 35 cm, although larger dimensions are better. The substrate should be a layer of sand at least 10 to 12 cm deep. Although light, fine-grained Sahara sand is ideal, this is almost impossible to obtain. Washed, sieved and dust-free river sand provides a suitable alternative. The vivarium should only be heated from above using spotlights or Elstein™ infrared heat lamps. Ground heaters, heating cables, or heating pads should never be used. Rocks and tree roots may be used for decoration. These skinks prefer temperatures between 37 and 42 °C to reach the desired body temperature, after which they will burrow down to the cooler layers of sand. During summer, the air temperature should be around 28 °C and the surface of the sand more than 35 °C, in places even as high as 60 °C. However, it should always be possible for the skinks to burrow to cooler levels.

To correspond to their natural habitat, these skinks require much light, such as that provided by high-pressure mercury vapour lamps. To be able to show their natural behaviour, these skinks require a light intensity similar to that in Central Europe on a hot, sunny July day, i.e. around 800 to 1,000 W/m^3.

The natural diet of these animals is half animal, half vegetable. They should be given meadow sweepings, mealworms, crickets, small locusts, seeds of sweet grasses, linseeds, and corn seeds (Hartmann, 1989). Although they can satisfy their low liquid requirement from their food, a bowl of fresh drinking water should be given each day.

The development of the male gonads depends especially upon the periodic variation of the light intensity. For around 9 months each year, the gonads are completely inactive. With favourable lighting, temperature, and feeding, the timing of the breeding season may be displaced, but only by a few weeks.

During the breeding season the males follow females constantly, trying over and over again to seize them in a copulatory bite to the neck. Finally, the male forces himself beneath the female from the side opposite to the copulatory bite, and mating takes place. According to Hartmann (1989), the actual copulation lasts from 1 to 4 minutes, but may be even shorter (Brüssing & Meyer, 1985).

During the breeding season, mating takes place almost daily if the light intensity is at least 800 W/m^3 and the temperature is 37 °C or higher. A female kept by Hartmann laid six soft-shelled eggs measuring from 21 to 24 × 10 to 12 mm. Unfortunately the entire clutch was infertile. The young should be reared on a mainly animal diet.

Sphenops Wagler, 1830
Wedge-Headed Skinks

These extremely smooth-scaled skinks live in North Africa and Israel in dry, sandy regions. They are closely related to the genus *Chalcides* and have a similar, cylindrical build. The snout is somewhat wedge-shaped and the lower jaw is very undershot. The slitlike nostrils are on the top of the head. The upper eyelid has a large "window" and the ear orifices are also slitlike. This skink has only very short, thin limbs and the number of toes may vary.

Sphenops sepsoides (Audouin, 1829)
Wedge-Headed Skink

Distribution: Egypt and Israel.
Description: In these 14 to 16 cm long animals, the forelegs are only half as long as the hindlegs. The digits are only a few millimetres long. A small number of pointed scales prevent sand penetrating into the minute ear orifice.
Habitat: Sand dunes with sparse vegetation.
Lifestyle, husbandry, and reproduction: Because these skinks are nocturnal they are rarely found in the wild. Attention is usually drawn to them by the tracks that they leave in the sand. When disturbed or threatened they dig even deeper into the sand, making it almost impossible to find them. The tail breaks off very easily and does not completely regenerate. The diet of this species consists of small insects and sweet fruit. In captivity they should be kept in the same way as *Chalcides* species.

Slender Skinks
Subfamily Lygosominae

Ablepharus Lichtenstein, 1823
Snake-Eyed Skink

The distribution range of the genus *Ablepharus* stretches from southern Europe to western Asia. There are four or five known species that reach lengths between 7 and 13.5 cm. These slender skinks live on bushy steppes and in light woodland. Their short, thin legs end in five toes. The lower eyelid is joined to the upper eyelid to form a "spectacle."

Ablepharus kitaibeli (Bibron & Bory, 1833)
Snake-Eyed Skink, European Copper Skink

Distribution: From Central Europe through the Balkan States to the Aegean and Asia Minor. In western Asia southwards to the Sinai Peninsula and northwards to Armenia.
Description: The Snake-Eyed Skink reaches a length of around 12 cm, of which two-thirds is the tail which runs straight down from the body and ends in a fine, elongated point. The brown to olive upper side has a beautiful sheen and appears almost metallic. The black sides are separated by a dark stripe that starts at the sides of the head, from the nostril above the eye, and ends at the tip of the tail.
Habitat: These skinks prefer areas with low, dry grass and areas with sparse vegetation. They are also found amongst piles of leaf litter and amongst small bushes or similar hiding places.
Lifestyle, husbandry, and reproduction: From midday onwards, these diurnal skinks hide beneath rocks or clumps of grass. They eat small insects and their larvae but

Ablepharus kitaibeli

will also take earthworms. In captivity they should be kept in the same way as *Chalcides* species. Mating begins after hibernation, lasting from October until April. Females lay two to four eggs in protected and secure places on the ground. The young, which measure around 33 mm, hatch after approximately 2 months. They should be kept in a separate vivarium and be fed on minute insects such as wingless fruitfly (*Drosophila*), springtails, and sifted meadow sweepings. They are best reared in a planted vivarium with some open spaces in which a large number of springtails have been placed. Part of the rearing vivarium should be kept constantly moist.

Acontias Cuvier, 1817
Lanceolate Skinks

The representatives of this genus live in Africa and Madagascar in dry, sandy areas. Their pointed snout indicates that this limbless, short-tailed lizard is a great burrower. Lanceolate Skinks have moveable, elongated eyelids and three to four brow scales over each eye. There is no ear orifice. The tail is short and stumpy. They are mainly active at night and eat the usual insects. In summer the female gives birth to between three and six young.

Acontias meleagris (Linnaeus, 1758)
Dark Lanceolate Skink

Distribution: Western and eastern Cape. Isolated populations occur along the Karoo.
Description: This species reaches a length of 20 to 26 cm; in exceptional cases it may

even reach a length of up to 30 cm. It has a slender head and body with a rounded snout. The eyelids are not transparent. Two subspecies are known. In the nominate form, the tail is not pointed. The back is unicoloured or may have scattered dark spots that form a crenated pattern. *Acontias m. orientalis* has a slightly pointed tail and six black stripes along the back. This subspecies is limited to the eastern part of the distribution range.

Habitat: These animals are usually found on dry, sandy ground, frequently beneath rocks or dead wood.

Lifestyle, husbandry, and reproduction: These skinks are excellent burrowers which eat all manner of invertebrates. They rarely drink water, but absorb moisture from their surroundings and food. Towards the end of summer the females give birth to up to four young measuring around 80 mm. In captivity this skink should be kept in the same way as *Chalcides ocellatus*, but with a substrate of at least 15 cm of granular sand. They should be hibernated at 15 to 20 °C.

Acontias percivali occidentalis Boettger, 1894
Western Lanceolate Skink

Distribution: In the north to Transvaal and Zimbabwe, westwards to Namibia, and southwards to Angola.
Description: Usually from 22 to 26 cm, maximum 30 cm long. These medium-sized, limbless skinks have a very slender body and head with a rounded snout. The lower eyelids are not transparent. The cylindrical tail is slightly pointed. The body is pale olive to grey-brown in colour.
Habitat: Bushland and savannahs.
Lifestyle, husbandry, and reproduction: See *A. meleagris*. These skinks are frequently found beneath dead and rotting wood where they search for insects and earthworms. In summer, females produce between one and three young.

Acontias plumbeus Bianconi, 1849
Giant Lanceolate Skink

Distribution: Natal and Transvaal.
Description: The Giant Lanceolate Skink, at up to 56 cm, is the largest member of this genus. It has a long snout, but no ear orifices on the wedge-shaped head. The head runs straight into the cylindrical body. The body is covered with smooth, gleaming brown or black scales that overlap in the manner of roof tiles. There are no limbs. The oval eyes and slightly downwards-pointing jawline give this skink an indolent appearance.
Habitat: Various habitats from desert sand through coastal sand dunes to forest floor.
Lifestyle, husbandry, and reproduction: In accordance with their size these skinks require a large vivarium similar to that described for the Sheltopusik (*Ophisaurus apodus*). In all other respects they should be cared for in the same way as *Chalcides*

Acontias percivali occidentalis

ocellatus, but without hibernation. For breeding however, a 4 to 6 week rest period at 15 to 20 °C is advisable. Apart from insects, these skinks will also eat earthworms. In April, 1974, a 52 cm female gave birth to seven young in captivity. In colour and appearance they were identical to the parent animal (Patterson & Bannister, 1988).

Cophoscincopus Mertens, 1934

The single species of this genus is *C. durus*, which is probably closely related to the genus *Panaspis*, but which has heavily keeled body scales.

Cophoscincopus durus (Cope, 1862)
Water Skink

Distribution: The mountains of West Africa; Guinea, Sierra Leone, Liberia, Ivory Coast and Ghana.
Description: A small brown skink with a total length of 11 to 12 cm. On the upper side, the keeled scales form longitudinal rows and run in oblique, angular rows. On the pink throat there are white and brown flecks. In males the underside of the body and front part of the tail are bright orange. On the back there is a black central stripe, bordered on each side by a black stripe broken at irregular intervals by yellowish-white spots. There are rows of white flecks on the flanks. The tail is speckled with light brown and white flecks.
Habitat: Common at altitudes between 700 and 900 m, on the banks of fast-flowing streams, beneath and amongst rocks, leaf litter, and driftwood (Mudrack, 1977, 1985).

Acontias plumbeus (above), Cophoscincopus durus (below)

Lifestyle, husbandry, and reproduction: Mudrack kept these skinks in an aqua vivarium without substrate. Flat, moss-covered rocks provided "islands." A clump of sphagnum moss served as refuge. The other furnishings were pieces of bogwood, moss-covered branches, and a small-leaved *Philodendron* plant.

The temperature was kept almost constant between 20 and 22 °C because it was obvious that the skinks were uncomfortable at higher temperatures. Food took the form of crickets, house flies, and small earthworms. After 2 months the female laid two 11 × 7 mm eggs amongst the sphagnum moss. The eggs were laid on a rock with a thick layer of moss and placed in a separate container. This container was filled with water to a depth of 3 cm and then covered with a sheet of glass. With relative humidity of almost 100% and temperatures between 20 and 22 °C, the 38 mm young, which were as thick as strands of wool, hatched after 49 days. For the first few days of life they were fed on springtails.

Dasia Gray, 1839
Tree Skinks

Earlier these skinks were assigned to the genus *Lygosoma*. The 16 presently known species live in the tropical forests of southeast Asia. According to species they reach lengths of 20 to 30 cm. The windowless eyes have moveable eyelids. The ear orifices are plainly visible. The well-developed limbs end in clawed toes, by means of which these lizards climb expertly on trees and bushes. Beneath the toes there are enlarged scales.

Dasia (Lamprolepis) smaragdina (Lesson, 1830)
Emerald Tree Skink

Distribution: Taiwan, Philippines, Palawan, and Sulu Archipelagoes, Micronesia (Palan Islands to the Marshall Islands and Farolinea), Indo-Australian Archipelago, Solomon and Santa Cruz Islands.
Description: The Emerald Tree Skink has a build typical of the family. The snout is fairly pointed with the head being barely set-off from the body. There are very sharp claws on the toes of these tree-dwelling skinks. The body may be various colours. Usually the head and front part of the body are emerald green, whilst the rear part, including the tail, is bronze. These animals are fully grown at 25 cm. The tail root of older males is thicker. The tail itself is conical down to the tip. In females the tail is reduced in size at the root where it remains around 1 cm thick; only behind this does it take on a conical shape (Manthey, 1985). In addition, in males, on the underside of the hindlegs at the ankle, a large, yellow, slightly convex scale may be found. In females, the scales at this joint are significantly smaller. Juveniles are unicoloured and without any transverse bands. This is normal for juveniles of the genus *Dasia*.
Habitat: Tree dwellers on flatland up to an altitude of 600 m. They are rarely found

on the ground and prefer bare tree trunks without climbing plants, in a position that receives much sunlight. They also live in mangrove forests, at the edges of rain forests, on individual trees, on coconut plantations, and even on minute islands with only three or four palm trees.

Lifestyle, husbandry, and reproduction: These skinks require a rain forest vivarium with a mixture of leaf litter and potting compost as substrate. Some gnarled branches and epiphyte stems will allow the animals to climb at will. The vivarium should be sprayed twice daily after which the animals will satisfy their liquid requirement by licking up the resulting water droplets that collect on the plants. During the day, a relative humidity of 50 to 60% is required. At night this should increase to 90 to 100%. It is advisable to place the vivarium in a sunny position. The temperature should vary between 25 and 27 °C during the day, localised beneath a spotlight to 35 to 40 °C. The vivarium should be brightly illuminated to allow the animals to exhibit their natural daytime activity. Stale air should be avoided at all costs. Tree Skinks feed on arthropods, woodlice, snails, and hairless caterpillars. They should also occasionally be given leaves, flowers, and sweet fruit.

Manthey (1985) found a vivarium of $66 \times 57 \times 70$ cm to be too small. At the slightest disturbance the skinks would panic and rush wildly around the vivarium before crashing into the front glass panel. This wild behaviour ceased only when the animals were placed in a larger vivarium measuring $90 \times 80 \times 150$ cm in which, when disturbed, the skinks fled to the upper reaches. In time, the skinks became almost tame. The height of the vivarium is a vital factor if these animals are to be housed correctly. These skinks should also be housed in pairs, since in the wild they will jointly defend their large territory. Very often they will also lead individual lives. If two skinks meet accidentally, it usually results in a very vicious fight (Peters, 1985).

Females lay up to six eggs beneath loose bark (Peters, 1985). On hatching, juveniles measure only around 25 mm. Manthey found a juvenile in his vivarium but was unable to determine when it had hatched or find the egg from which it had emerged. This juvenile was placed in a small rearing container and kept at a maximum temperature of 25 °C. Because the juvenile was not feeding, the temperature was increased so that the lower area was at 26 to 27 °C, whilst the upper reaches were at 35.5 to 39 °C. The juvenile then began to feed well and became as agile and active as the parents. Within only 1 month, the rearing container proved to be too small, and during the following months the juvenile had to be moved to ever larger vivaria.

Dasia olivacea Gray, 1838
Olive Tree Skink

Distribution: Southeast Asia, south of 15° northern latitude; Burma, Thailand (Yale, Narathiwat, Pattani, Satun), Cambodia, Laos, Vietnam, Malaysia, Borneo, Java, Sumatra, Natuna Islands, Andaman Islands, and Nicobar Islands.

Dasia smaragdina (above), Dasia olivacea (below)

Mabuya capensis

is tail. The head is broad with a slightly compressed snout. The stout body has four particularly well-developed limbs. The ground colour of the back and flanks is brown, occasionally olive-brown. There are three longitudinal stripes along the back, and on the underside a stripe that is slightly broader than the stripes on the sides. Along the body and tail there are black bars that are broken by three longitudinal stripes and black-bordered, white eyespots. The underside is greyish white to white.
Habitat: These skinks prefer areas of dense vegetation but may also often be found in open spaces and dry sandy areas.
Lifestyle, husbandry, and reproduction: Matz (1971) kept these Mabuyas together with *Chalcides ocellatus* in a dry vivarium. At first they were extremely timid, but quickly became more trusting. The substrate should be a layer of sand upon which pieces of bark and several rocks should be placed to provide refuges. Drinking water is essential.

Beneath a spotlight the temperature should be between 35 and 38 °C. They will eat all manner of insects such as grasshoppers, crickets, caterpillars, mealworms, beetles, and butterflies. Because males are extremely aggressive towards one another, this is another species that should only be kept in pairs. Females usually give birth to 9 or 10 young. In Rose's case (quoted in Matz), in January a female produced 15 young, each measuring around 7.5 cm. Two, or at most three, young should be reared together and apart from the parent animals.

Mabuya carinata (Schneider, 1801)
Keeled Skink

Distribution: India, Sri Lanka.
Description: This 28 to 30 cm skink has very delicate limbs and its tail ends in a fine point. On the glossy brown to olive-brown back there are black flecks and dashes. From each eye to the tail there is a light-colored side stripe. The underside is white, or sometimes light yellow. During the breeding season, males have scarlet sides and a yellow belly.
Habitat: Mainly in moist woodland and cultivated areas with dense vegetation.
Lifestyle, husbandry, and reproduction: These mainly ground dwellers are also excellent climbers and therefore require a spacious vivarium. A suitable substrate is a mixture of leaf litter and peat. A tree root placed in the vivarium will provide a refuge and will retain moisture after the daily spraying. Whenever possible, these skinks should be exposed to direct sunlight. The diurnal Keeled Skink requires temperatures of 26 to 29 °C during the day and 22 °C at night. They may be fed on all manner of arthropods. They can be housed with other *Mabuya* species, provided the vivarium is large enough. Because males can be aggressive towards one another it is wise to keep this species in pairs or one male with two females. Females lay eggs. Further details may be found under *Mabuya multifasciata*.

Mabuya macularia (Blyth, 1853)
Flecked Mabuya

Distribution: From Pakistan through India and Sri Lanka to Indochina, southern Thailand, and the northern part of western Malaysia.
Description: There are several races of *Mabuya macularia* but for a layperson they are extremely difficult to distinguish. Usually those species available commercially are from southern Thailand. These skinks have a SVL of 6.5 cm and a tail length of 9.0 cm. The top of the head is olive-brown. The similarly coloured body has numerous dark flecks, which on the front of the body form four longitudinal stripes. Males also have white flecks which increase in number towards the rear of the body. There is a light dorsal stripe down the tail. Below this there is a dark stripe with scattered light flecks. Below this stripe there is another light stripe that is either continuous or consists of spots. The throat of males is orange-red. A further indication of sex is the greatly enlarged hemipenes pockets of males. The dorsal and side scales have five or more rows of keels. Juveniles are similar in colour to adults.

Habitat: This species prefers sunny places in both rain forests and cultivated areas up to altitudes of 0 to 600 m.
Lifestyle, husbandry, and reproduction: These small and undemanding skinks may be kept in pairs in vivaria as small as 80 × 40 × 50 cm. Temperature and furnishings are as described for *Mabuya multifasciata*. They eat meadow sweepings, insects, and small spiders. If fed a varied diet they do not need large supplements of calcium and vitamins. According to their origins, the breeding seasons and number of eggs produced may vary considerably. Females lay two to six hard-shelled eggs measuring 11 × 9 mm. These are either buried or laid in rotting vegetation or beneath fallen trees. The eggs should be artificially incubated at 27 to 29 °C, never below 25 °C. The young have a SVL of around 20 mm with the tail being some 25 mm long. Ideal foods for the young are meadow sweepings, minute crickets, and waxmoth larvae.

Mabuya multifasciata (Kuhl, 1820)
Four-Striped Mabuya

Distribution: From northern India (Assam) eastwards to Yunnan and the island of Hainan, southwards through Burma, Thailand, the Andamans, and Nicobars, western Malaysia and the entire Indo-Australian Archipelago including the Philippines to New Guinea.
Description: These Mabuyas may reach a SVL of 12.5 cm with a tail of 18 cm. Along the back there are usually three rows of significantly keeled scales. The scales on the sides may be smooth or keeled. *Mabuya multifasciata* varies not only geographically, but also individually. The back is brown to olive-brown, unicoloured or with almost-black longitudinal rows of individual spots. The flanks are dark brown to black, especially at the top, and may also have some black-bordered white flecks. In both sexes the flanks may even be red to yellow. The underside is usually white. Some specimens have a yellow throat and chest.
Habitat: As a ground dweller, in primary and secondary forests, on cultivated land, and even in the vicinity of human habitation. More common on flat land, but also found up to an altitude of 2,000 m.
Lifestyle, husbandry, and reproduction: These Mabuyas seek out sunny resting places. When searching for food they hunt in pairs amongst rocks, on bushes, on trees, and on fence posts. When disturbed they will quickly disappear into the leaf litter or decaying tree trunks. They always live near water. They are adept swimmers and divers. This is not a particular species as far as food is concerned. They will eat anything that moves and that they can overpower, even smaller lizards. Insect pupae also form part of their diet. During the breeding season males indulge in serious fighting. Females produce four to eight live young with a total length of around 81 mm. In some parts of the distribution range these skinks breed throughout the year.
    A vivarium of 100 × 50 × 50 cm is required for a pair. During the day, one-third of the vivarium should be heated in such a way that the ground reaches 33 to 40 °C

Mabuya macularia (above), Mabuya multifasciata (below)

and is brightly lit. The remainder of the vivarium should be cooler and more moist. During the day, the Four-Striped Mabuya requires a temperature of 25 to 27 °C and humidity of 70 to 80%. At night the temperature should fall to 20 to 25 °C, which increases the humidity. A water section is not absolutely vital, but is recommended. A suitable substrate is a mixture of loam and leaf litter upon which rocks, a tree root, and pieces of bark should be placed to provide refuges. If desired, the vivarium may be planted. Cacti and other succulents are, however, unsuitable. Depending upon the variety of plants in the vivarium, it should be sprayed every 2 or 3 days. Food, in the form of all manner of arthropods, may be given. Moreover, sweet fruit will also be eaten occasionally. It is not wise to house these skinks with smaller lizards—these would invariably be quickly devoured. Likewise, the young should be separated from the parents immediately after birth, and reared in individual, small vivaria. Rearing these skinks presents no problems. On the contrary, this is an excellent species for a beginner. A suitable rearing container is a smaller version of the vivarium described for the adult skinks.

Mabuya quinquetaeniata (Lichtenstein, 1823)
Blue-Tailed Skink, Five-Lined Skink

Distribution: Northeast to southeast Africa, south of the Sahara Desert and over the entire Nile Valley.
Description: These skinks can reach a total length of 25 to 30 cm. In females and juveniles, the black upper side has cream longitudinal stripes. The tail is blue and the belly dirty-white. Males have a brown dorsal surface upon which there are two broad reddish brown stripes and a narrower light brown dorsal stripe. The upper side of the tail in males is brown to red, whilst the underside and belly are a yellowish colour.
Habitat: Savannahs and steppes. As followers of civilisation, they frequently live in the vicinity of towns and villages.
Lifestyle, husbandry, and reproduction: Blue-Tailed Skinks are very aggressive towards other lizards (Laurens, 1976) and require a dry vivarium with temperatures between 25 and 30 °C, localised beneath a spotlight to around 45 °C. At night the temperature should fall to 18 to 22 °C. A heating pad or heating cable will help to achieve the desired temperatures. A suitable substrate is sharp sand, part of which should be kept permanently moist. Rocks, branches, and tree roots allow the skinks to climb and serve as refuges. A shallow bowl of drinking water should be available at all times.

In the wild these skinks are very sociable. A pair with several young will hold and defend a large territory. The 11 to 15 mm long eggs are laid in a moist place in the vivarium and are carefully covered. After an incubation of 35 to 40 days at temperatures between 27 and 29 °C, the young hatch measuring from 6 to 8 cm in length. They should be kept in the same way as other Mabuyas.

On October 20, 1974, a female kept by Laurens buried eleven dirty-white eggs in a flower pot containing an *Agave* plant. These eggs were 15 mm long and were carefully covered. The eggs had been well concealed between the roots of the plant, and it was there that they developed because the flower pot was constantly moist.

The eggs were eventually transfered to an incubator. With an average temperature of 29 °C and high humidity, the first slits in the eggs were noticed on November 25, 1974. The young remained inside the eggs for a further 24 hours. When they finally emerged, the young, which had magnificently coloured tails, were immediately transferred to a smaller vivarium for rearing. This vivarium measured 35 × 25 × 25 cm and contained several pieces of cork bark. After only 12 hours the young ate small crickets and waxmoth larvae that had been previously dusted with calcium and vitamin powders. Vitamins were also added to the drinking water. The young drank in the morning and enjoyed basking beneath a 25-W incandescent lamp, below which the temperature varied between 40 and 45 °C. There were no indications of aggression amongst the young. They were however very excitable, twitching their tails at the slightest disturbance.

On December 2, the female laid an additional 10 eggs in the same place. The young hatched from these eggs 5 weeks later. In that time, the young from the first clutch had grown to a length of 15 cm. A third clutch was laid on January 18, 1985, again in the same flower pot. Because of a shortage of calcium the shells were much weaker and became damaged as the eggs expanded. After the eggs had "sweat," the first three young hatched on March 4. Two further young hatched on March 6. Each of these young had a total length of 8 cm.

Mabuya vittata (Gmelin, 1789)
Striped Mabuya

Distribution: From eastern Algeria to Egypt and from Israel through Syria to Asia Minor.
Description: The Striped Mabuya can reach a total length of over 20 cm of which the tail is a good 20%. The ground colour is brown. Four rows of dark flecks begin on the head and run down to the tail joint.

A broad yellowish stripe starts from each side of the snout and runs along each flank to the hindlegs. The ventral surface is greyish white. The young are uniform brown without markings.
Habitat: Grass and bushland, forest edges up to an altitude of 200 m.
Lifestyle, husbandry, and reproduction: In the wild, the diet consists mainly of insects, spiders, and young lizards should the opportunity arise. These Mabuyas drink from puddles after rainfall (Esterbauer, 1985). They also satisfy their liquid requirement from their food or by licking up droplets of dew in the early hours of the morning.

In Syria, Esterbauer kept several Mabuyas in an outdoor enclosure measuring 120 × 50 × 55 cm. Each of the sides, apart from the front glass panel was covered in fine gauze. The substrate was a layer of sand 6 cm deep. The enclosure contained climbing branches and a long, narrow rock that had many holes, and a water bowl.

The Mabuyas climbed happily around the enclosure and spent long periods basking. In the evening they burrowed beneath the rock. They remained there overnight and during overcast weather. In September, the female gave birth to two young, one of which was obviously unable to free itself from the amniotic sac and subsequently died. The living young had a SVL of 29 mm and a tail length of 42 mm. In captivity, these Mabuyas will eat crickets, grasshoppers, mealworms, beetles,

Mabuya vittata

waxmoths, and almost any other insects that are presented to them. In an indoor vivarium the temperature should be between 25 and 30 °C during the day, localised beneath a spotlight to around 40 °C. At night the temperature should fall significantly.

Mochlus Günther, 1864

Earlier, some of these skinks were assigned to different genera. The lower eyelid has scales. There is no "window." The limbs are very powerful. If they are laid towards the centre of the body they do not touch. *Mochlus* species always have five fingers and toes.

Mochlus fernandi (Burton, 1836)
Magnificent Skink, Fernand's Slender Skink, Fernand's Magnificent Skink

Distribution: Africa. Fernando Po, Cameroon, Gabon, Angola, Togo, and Liberia.
Description: These beautiful skinks reach a total length of around 35 cm. The back is light brown to medium brown whilst the flanks are red. Along the flanks, white spots form transverse bands. From the forelegs to the head, the black spots run together to form a large black patch that is broken by individual white scales. The up-

Mochlus fernandi

per jaw, below the eye, is red. The throat also has white spots that are frequently arranged in rows. The limbs are dark brown, in juveniles, almost black.

Habitat: Tropical rain forest.

Lifestyle, husbandry, and reproduction: These breathtaking creatures are predominantly ground dwellers that are active during the hours of daylight. Mudrack (1976) kept three specimens in a 80 × 45 × 45 cm vivarium with a 10 cm layer of potting compost as substrate. The vivarium contained blocks of peat beneath which the skinks burrowed. *Scindapsus aureus* and *Hoya carnosa* were planted in the vivarium to provide decoration. A tree root allowed the animals to climb and provided further decoration and hiding places. The vivarium was lightly sprayed daily. The skinks were fed on crickets, slugs, mealworms, and strips of meat and fish which were taken from forceps.

Unknown to their keeper, the female laid nine eggs from which eight young hatched in the vivarium. The parchment-like eggs measured some 25 × 15 mm. The young were transferred to a separate vivarium of 40 × 20 × 30 mm which also had a 10 cm layer of potting compost as substrate. They fed avidly on crickets and earthworms. After only 6 months they had reached a length of 15 to 16 cm, at which time they were placed in a larger vivarium. Because it was almost impossible to keep a check on each animal, it was not noticed that they were suffering from calcium deficiency. They subsequently had to be euthanised. The adults produced two further clutches of young, one of eight and the other of five. By putting calcium in their food and vitamins in their drinking water, deficiencies were completely avoided.

Mochlus sundevalli Smith, 1849
Slender Skink, Sundevall Magnificent Skink

Distribution: Northern South Africa. The northernmost distribution limit is southern Angola and Somalia. The southernmost limit is the Kalahari-Gunsbok National Park. The subspecies *M. s. somalica* occurs only in Somalia and is characterised by the long fifth toe.
Description: At only 15 to 19 cm these are very small skinks. They have well developed limbs, each with five toes and a stout, smooth tail. The eyes have scaled, moveable eyelids. The ear orifices are very small and deeply inset. The smooth-scaled, light to dark brown body usually has a dark spot at the base of each scale, creating a fine speckled pattern. The underside is creamy white and may have speckles on the underside of the tail.
Habitat: Dry sandy areas and dry hillsides.
Lifestyle, husbandry, and reproduction: The Slender Skink lives on sandy ground amongst old rocks and wood. Its diet consists mainly of small arthropods. In summer, females lay between two and six eggs which they bury in the ground. Eggs are frequently also laid in termite mounds. Hatchlings measure 50 to 55 mm. These skinks should be kept in the same way as *Chalcides ocellatus*.

Riopa Gray, 1839
Riopa Skink

Riopa species live in dry steppe-like regions. The largest of these may grow to a length of 35 cm. Some species of these slender, elongated skinks have lightly keeled scales. The lower eyelid has a partially transparent "window." Their limbs are short and each has five toes.

Riope koratense (Smith, 1917)
Khorat Skink

Distribution: Thailand.
Description: The 20 cm Khorat Skink has a cylindrical body and only very small ear orifices. The upper side is brown, reddish brown, or reddish violet. The sides are often yellowish green. The underside is yellowish white with dark speckling limited to the tail. The tail root is very thick.
Habitat: Low limestone hills with many refuges in the form of clefts and rock fissures.
Lifestyle, husbandry, and reproduction: The Khorat Skink often has scars from bites because they are very intolerant and quarrelsome with one another. They are constantly burrowing in the substrate, which should be a 5 cm layer of peat and sand

that should be kept permanently moist, enabling the animals to dig their burrows more easily. Plants placed in the vivarium must remain in their pots. During the day the temperature should be 23 to 26 °C, localised in places to 30 °C. This will allow the skinks to seek out the most comfortable place at any time. At night the temperature may be allowed to fall to around 20 °C.

In captivity, these skinks will feed on earthworms, crickets, mealworms, waxmoth larvae, small beetles, woodlice, centipedes, and similar animal matter.

The Khorat Skink is normally only seen in the evening when they begin looking for something edible. They usually devour their food underground.

Two females kept by Harbig (1988) increased dramatically in girth around mid-February and, on February 26, each laid three eggs in a plant pot. One of the two young that hatched on April 27, measuring 40 mm, died. The other reached a total length of 72 mm at only 100 days old.

Scelotes Fitzinger, 1826
Scelotes Skinks

The skinks of this genus live in Africa, south of the Sahara Desert, and on Madagascar in fairly moist regions. They are very elongated and may reach a length of up to 50 cm. Some species have complete five-toed legs, whilst other species are completely without limbs. The body is covered in small, smooth scales. In species with fully developed limbs, the tail is longer than the body. In species without limbs, the tail is much shorter than the body. External ear orifices may be present or absent.

Scelotes species are avid burrowers that live amongst leaf litter or on grassland. Some species live near water where they feed on water insects and fish. They will, however, also feed on land insects. They are ovoviviparous.

Scelotes bipes (Linnaeus, 1758)
Two-Toed Scelotes Skink

Distribution: The nominate form can be found from the Mossel Bight to the Cape Peninsula. The subspecies *S. bipes sexlineatus* is found from Clanwilliam northwards to Port Nolloth.
Description: A small skink, usually around 12 to 14 cm in length. The stumpy hindlegs end in only two toes. The head appears slightly compressed and the lower eyelids have scales. The ear orifices are very small. Along the centre of the body there are 18 rows of scales. The tail is shorter than the body, the upper side of which is silver grey to pale brown. The scales are dark at the centre creating a pattern of spots or stripes. The lighter underside has pale flecks.
Habitat: Sandy, coastal areas.
Lifestyle, husbandry, and reproduction: These skinks are usually found beneath rocks or burrowed into the sand. They feed on all matter of insects, spiders, worms, and other invertebrates. In March, females give birth to two live young. In captivity they should be kept in the same way as *Chalcides ocellatus*.

Riopa koratense (above), Scelotes capensis (below)

Scelotes capensis (Smith, 1849)
Cape Scelotes

Distribution: Namaqualand to Central Namibia.
Description: These small 9 to 10 cm skinks have fully developed forelegs and hindlegs, each with five toes. They very rarely reach the maximum length of 12 cm. The

head is very compressed and the lower eyelid is transparent. The ear orifices are small and circular. There are 22 rows of scales at the centre of the body. The tail is slightly longer than the SVL. On the upper side, these skinks are light olive to olive-brown with a copper-coloured sheen. The tail is brown to black. A pale olive-yellow to yellowish brown stripe runs along both sides of the back from the head to the tail where it frequently changes to a blue or blue-grey colour. The underside is greenish yellow to pale yellowish brown. The throat and chest are somewhat paler.
Habitat: Rocky areas rich in succulent vegetation.
Lifestyle, husbandry, and reproduction: This species leads a very secretive lifestyle beneath rocks on sandy ground, often beneath succulents or dead plants. This species may also be kept in the same way as *Chalcides ocellatus*.

Tribolonotus Duméril & Bibron, 1839
Helmeted Skink

Helmeted Skinks live in New Guinea, the Solomon Islands, and in New Caledonia in moist, shaded valleys in the vicinity of water. Some species may reach a total length of 20 cm. The head is elongated and has the appearance of a helmet. The scales on the neck are especially spiny. The large, heavily keeled scales are very prickly. This is the only lizard with glands below the ventral scales and on the flat of the hands and feet. In females, the left ovary and oviduct are stunted. In general, these skinks are uniform dark brown in colour.

Tribolonotus novaeguineae (Schlegel, 1834)
Bush Crocodile, New Guinea Helmeted Skink

Distribution: New Guinea.
Description: The back and tail of these skinks has four longitudinal rows of spines. The scales on the legs are always pointed. The crude head with hard scales is particularly striking. The width of the head increases sharply towards the neck where it cuts back to a straight line. Around each eye there is a yellow circle and at the corner of each eye an orange spot. These skinks are dark brown in colour. They reach a SVL of 80 mm and a tail length of 82 mm.
Habitat: Peters (1985) found these skinks in dense jungle amongst coconut palms, beneath moss-covered, fallen trees, or beneath piles of decaying coconut fibre. The temperature here was almost constant at around 24 °C and the humidity was 95 to 100%.
Lifestyle, husbandry, and reproduction: In the vivarium, these nocturnal creatures only appear when the lighting has been switched off. The substrate should be a mixture of sand and peat around 20 cm deep. Suitable plants are *Philodendron, Monstera, Bilbergia,* and *Aechmaea*. The substrate should form a small mound at the centre of the vivarium. It is here that the animals will build their home. If a flat rock is placed on this mound, the skinks will excavate a tunnel and dwelling chamber be-

neath it. When molested, these skinks emit a loud croak. Nothing is yet known about their breeding habits in captivity.

NIGHT LIZARDS
Family Xantusiidae

The family of Night Lizards contains four genera with a total of only 12 species. Their systematic position is still the subject of considerable debate amongst herpetologists and the solution may be found in taking a closer look at the Family Gekkonidae. The representatives of the original family live in dry areas, along tropical coasts or in pine forests of the Central American Mountains. In appearance, Night Lizards are somewhat reminiscent of the European Common Lizard *(Lacerta vivipara)*. The head is very blunt and has large scales. The lower, transparent eyelid is attached to the upper edge of the eye forming a lens which these lizards—like geckoes—keep clean by licking. The dorsal scales, at least those in the centre, are small, flat, and smooth, whilst the ventral scales are large and rectangular. The tail is cylindrical and is usually equal to the SVL. It regenerates well. Males have femoral pores.

The construction of the spine and some of the features of the skull are similar to geckoes, but the two groups differ in the construction of the temporal arch.

During the day, Night Lizards conceal themselves beneath rocks, in rock crevices, amongst moss-covered wood, or beneath leaf litter. At night they emerge to hunt for insects. As far as is currently known, Night Lizards are ovoviviparous and usually give birth to two, rarely one or more than two, young. Various species of Night Lizard are classed as "easy to keep" and are frequently bred in captivity. The captive husbandry conditions, however, vary according to the origins of the animals.

In 1989, Glaser reported that Night Lizards are not necessarily nocturnal creatures because they have a well-developed parietal eye. In laboratory tests, *Xantusia v. vigilis, Xantusia v. arizonae* and *Xantusia henshawi* were found to be able to tolerate a temperature of 0 °C and that at only 5 °C were capable of movement. It is therefore obvious that during the day their body temperature is regulated by basking.

Cricosaura Grundlach & Peters, 1863
Cuban Night Lizard

The solitary species of this genus lives in the tropical forests of Cuba and has several characteristics, i.e., two nasal scales, an unpaired cranial scale, and no frontal scale, which would make the creation of a separate subfamily Cricosaurinae rather obvious.

Lizards

Cricosaura typica Grundlach & Peters, 1863
Cuban Night Lizard

Distribution: The hinterlands of the southwest Cuban province of Oriente (Cabo Cruz) in an area of only 200 km² (Petzold, 1969).
Description: A slender, short-legged, long-tailed, brown lizard with a SVL of only 3 to 4 cm and a tail of 6 to 8 cm. The head, which is broad at the back, tapers from the ear orifices to the snout. The pupils are vertically elliptical.
Habitat: According to Petzold, these lizards live in the tropical regions of the island. He found these creatures in rocky forests, usually hidden beneath rocks.
Lifestyle, husbandry, and reproduction: This is a ground-dwelling species that is exclusively active at night. Very little is known about its lifestyle. Petzold kept a single specimen in an all-glass vivarium. The substrate was a layer of a sand/peat/leaf litter mixture upon which pieces of bark, rocks, and a shallow drinking bowl were placed. The vivarium also contained a number of *Tradescantia* creepers. The temperature in the vivarium varied between 25 and 35 °C and the humidity was kept at a medium level. If the substrate became too dry, the lizard would lie in the water bowl, with head erect, for up to 20 minutes. This nocturnal lizard ate small field crickets and house crickets, minute beetles, and meadow sweepings. During the day, these lizards usually remain buried in loose ground. At night the temperature in the vivarium should fall to around 5 to 6 °C.

Xantusia Baird, 1859
Night Lizards

These relatively flat Night Lizards have a broad head. Three species of this genus are known. The dorsal scales are small and smooth. The femoral pores of the males are larger than those of the females. These animals are capable of a physiological colour change from light to dark. During the daytime rest period they usually assume dark colouration.

Xantusia henshawi Stejneger, 1893
Granite Lizard, Granite Night Lizard

Distribution: Southern California southwards into northern Baja California (Mexico), and from there into the regions of the Sierra San Pedro Martir, up to an altitude of 2,300 m.
Description: Granite Night Lizards reach a total length of up to 14.5 cm. On the underside of the thighs, males have a white oval spot in front of the row of femoral pores. The strongly compressed body has dark brown to black flecks that expand during the day and contract at night. The skin is very soft to the touch, The large scales on the head are symmetrically arranged. On the underside there are large, smooth, angular scales.

Xantusia henshawi

Habitat: Deserts and semi-deserts, rocky gorges, and mountainsides, mainly on shaded, rocky slopes and in the vicinity of water.

Lifestyle, husbandry, and reproduction: According to Switak (1978), these lizards live in rock crevices and similar refuges and only rarely venture into the sunlight. Narrow rock crevices would appear to give them a sense of security, making it essential that stacked rocks are placed in the vivarium. The rocks must be arranged in such a way that the crevices may be inspected from the outside. Gravel of various grades makes an ideal substrate. During the day, the temperature must not exceed 25 °C, and must fall at night to 13 to 15 °C. A shallow bowl of clean drinking water is vital. The Granite Night Lizard will eat all manner of insects. Each year females give birth to one or two live young which are identical in appearance to their parents.

Xantusia riversiana (Cope, 1884)
Island Night Lizard

Distribution: Off the coast of California on islands such as Santa Catalina.
Description: The Island Night Lizard reaches a total length of 20 cm. They have a very sturdy build, a high, strong head with strong jaws and teeth. The small, slitlike

pupils expand in the dark. On the throat and sides of the body there are the typical folds of skin. The ground colour is dark grey with a light flecked or lattice pattern. The underside is pale grey, sometimes with a blue sheen that becomes yellow towards the tail.

Habitat: In rocky areas with little grass and many cacti.

Lifestyle, husbandry, and reproduction: During the day these lizards remain hidden beneath rocks, dead wood, or other cover. They are less timid than other Night Lizards and are occasionally seen during the day. Peters (1965) obtained 13 specimens between 6 and 16 cm long, all of which had regenerated tails. The stomach of one dead specimen contained the remains of wing cases of hard-shelled beetles, wings of dragonflies, seeds of grasses, and blades of grass. Thirty percent of the diet of these lizards consists of vegetable matter.

In accordance with the conditions where the lizards were found, Peters furnished the vivarium with a gravel substrate, leaf litter, and some large rocks. Four lizards were placed in this vivarium, but they were rarely seen. They were fed (in the United States) on mealworms, beetles, small scorpions, moths, crickets, lettuce, and pieces of tomato. During the day the temperature was 25 to 28 °C, falling at night to 22 to 24 °C. In the natural habitat, the temperatures are often much lower.

Xantusia vigilis Baird, 1858
Desert Night Lizard, Yucca Night Lizard

Distribution: The southwest of the United States, Baja California and Mexico, Mojave Desert, deserts in western California, in southern Nevada southwards to Utah and the northwest parts of Arizona.

Description: These 9.5 to 12.8 cm lizards are olive, grey, or dark brown. The body has numerous black spots. From the eye to the shoulder there is usually a light stripe with a dark edge. The skin on the back is fine and granular. The slitlike pupils are vertical. There are no eyelids. Males have large femoral pores, giving the thigh, in cross section, a triangular shape with rounded edges.

Habitat: Deserts, amongst fallen Joshua trees (*Yucca brevifolia*) or other plants of the *Yucca* genus.

Lifestyle, husbandry, and reproduction: A small vivarium with a ground area of 50 × 30 cm is adequate for a breeding group. A suitable substrate is a mixture of sand and gravel to which a little leaf litter has been added. The most suitable furnishings are *Yucca* stems that have been hollowed out by insects. In the absence of these, hollow branches are a good substitute. These should be placed beneath the heat and light source. During the day the temperature should be around 30 °C. At night 15 to 20 °C is sufficient. These animals are not only active at night, but may also frequently be seen during the day, especially at dusk. It is only on very hot summer days that these "Night Lizards" become fully nocturnal. It is thought that the parie-

Xantusia vigilis

tal eye acts as a form of "heat meter" to determine the extent and duration of activity.

If the parietal eye is covered by a small piece of aluminium foil, the activity increases. As reported by Glaser (1990), the daily cycle may vary considerably.

Night Lizards must always have fresh water and will feed on small spiders, house crickets, small grasshoppers, and other arthropods which are well chewed before swallowing. A breeding group may consist of two males and up to six females, each of which may give birth to one to three fully developed young each year. The young are much darker in colour than the adults and are easy to rear when fed on meadow sweepings.

Cowles (quoted in Klingelhöfer, 1957) reported on the birth of *Xantusia vigilis*. Around 20 to 25 minutes before the birth, the normally placid female became extremely restless, moving backwards and pushing upwards with all four legs. The first of the young quickly appeared, surrounded by an amniotic sac which the mother ruptured before the baby had left her body completely. The baby immediately began to struggle. When it stopped moving, the mother prodded it gently. When it finally left the amniotic sac it ran a few steps. At birth, the young are around 5 cm long. Cowles kept these lizards at 22 to 31 °C, usually around 26.5 °C. They were fed mainly on termites. Gravidity lasts for around 4 months.

TEIIDS
Family Teiidae

America is the home of Teiid Lizards. There are around 40 genera with approximately 200 species in the most divergent habitats from rain forests to deserts, even up to altitudes of 4,000 m. These lizards are considered to be the New World relatives of the European Lacertids. Despite the spatial separation of both families, there are very few differences in build and appearance. In the case of Teiids, the scales on the head are not attached to the bones of the skull.

The multiplicity of shapes and forms is very impressive. There are species that are only 10 cm long and others that measure over 1 m, having the appearance of a Monitor Lizard. Some of the smaller species are reminiscent of skinks but have bony skin.

Normally the body is covered in granular and overlapping scales of various sizes that may form regular, transverse, longitudinal, or angular rows. The large scales on the underside are particularly noticeable. In some ground-dwelling species there are fused, transparent eyelids. These lizards usually have four well-developed limbs, although there are species in which the limbs are only rudimentary. Their food consists essentially of insects and other arthropods, but there are larger species which also eat small mammals. Only very few species are completely vegetarian. Above all, there are many species about which relatively little is still known because, in the science of herpetology, they still play only a minor role.

Ameiva Meyer, 1795
Ameivas

The Ameivas live in a range stretching from southern Mexico to tropical South America, including the Antilles. They have been introduced into several areas of Florida. There are approximately 14 species that live in light forests and on cultivated steppes. They reach a total length of 15 to 55 cm. The tail is extremely long and the snout sharply pointed. The lower eyelid is semitransparent. The long tongue can be withdrawn into a tubelike sheath. The dorsal scales are granular whilst those on the underside are large and smooth, arranged in 6 to 16 longitudinal rows. The powerful limbs end in sharp claws.

Ameiva ameiva (Linnaeus, 1758)
Giant Ameiva

Distribution: Southern Mexico to Uruguay; introduced to Florida.
Description: These Ameivas reach a total length of 63.5 cm and are blue to brown in colour. Along the back there are rows of light flecks. Occasionally there is a pale

Ameiva ameiva

brown stripe along the centre of the back. On the underside there are 10 or more longitudinal rows of large, rectangular scales.
Habitat: Open, sunny, grassy landscapes, usually on sandy ground.
Lifestyle, husbandry, and reproduction: As ground dwellers, Ameivas enjoy digging and are active throughout the day. They therefore require a very large vivarium with a deep substrate of sand and loam. Half of the vivarium should be kept constantly moist, but not too wet. During the day these lizards require a temperature of 25 to 30 °C, falling at night to around 20 °C. Ameivas should be kept in pairs because they can be extremely aggressive. They will eat all manner of arthropods, and will also take nestling mice and occasionally overripe fruit. The furnishings should consist of tree roots beneath which the lizards will find refuge and sleep. Potted plants may be installed in the moist part of the vivarium. The most suitable plants are succulents or other broad-leaved, robust species. The vivarium should be sprayed daily and a drinking bowl is essential. As well as bright lighting, these lizards should also be given ultraviolet light weekly.

Depending upon age and size, females lay between one and four eggs. In some regions, oviposition depends upon the season, whilst in others, breeding takes place throughout the year. Occasionally these lizards should be subjected to "bad weather periods" at lower temperatures.

Callopistes Gravenhorst, 1838

These Tegus live west of the Andes from Equador to Central Chile in stony deserts and steppelike regions. On the Pacific side of the Andes, they are even found in the mountains. These lizards have powerful limbs. Dorsal scales are small and granular whilst the ventral scales are large and smooth.

Callopistes maculatus

Callopistes maculatus Gravenhorst, 1838
Chilean Tegu

Distribution: North and central Chile.
Description: These typical Tegus are vivid brown to reddish brown with white-bordered dark flecks. They can reach a total length of up to 50 cm.
Habitat: Dry areas on the western slopes of the Andes.
Lifestyle, husbandry, and reproduction: Although these animals are frequently available commercially, very little is known about their lifestyle. In the south of their distribution range, these Tegus hibernate for several months, during which time they live from the reserves of fat contained in the tail. Because of their size, they should be kept in the largest possible vivarium with a firmly cemented pile of rocks forming the centrepiece. The rocks should be cemented together in such a way that an almost natural cave is formed. It is here that the lizards will seek refuge and sleep. Because this is a burrowing species the substrate should be a layer of fine gravel 5 to 10 cm deep, part of which is kept permanently moist. A bowl of fresh drinking water must always be available.

During the day temperatures of 25 to 35 °C are necessary whilst at night they may be allowed to fall to 15 to 20 °C. The highest temperature should be beneath a spotlight. These Tegus eat all manner of large insects but will also eat nestling mice or other small mammals. Tegus of this species will thrive if given vitamins, calcium, and ultraviolet light weekly.

From early summer until late autumn, the Chilean Tegu may be placed in an outdoor enclosure, provided most of it is covered by rigid, transparent sheeting. This sheeting protects the animals from adverse weather whilst allowing the penetration of ultraviolet light. Dry greenhouses are especially suitable for these lizards.

In autumn the Tegus should be returned to their indoor vivarium in which the temperatures have been reduced. After the animals have fasted for 2 weeks with light daily spraying, the vivarium should be placed in a dark place at 10 °C to 15 °C for approximately 2 months. During this time however, the vivarium should be lightly sprayed every 3 days. The Tegus will hibernate during this time.

After this hibernation, the temperatures should be gradually returned to normal and the lizards given a rich and varied diet. Depending upon the space available, these lizards should be housed either separately or in pairs. After the hibernation it is wise to house them in pairs. It is not unknown for these Tegus to come into breeding condition whilst receiving ultraviolet light.

Cnemidophorus Wagler, 1830
Whiptail Lizards

Around 40 species of this genus live within the distribution range, which extends from the United States to northern Argentina. In this range they inhabit the most divergent habitats. The tail is around twice the SVL. Larger species may reach a total length of up to 35 cm. The dorsal scales are small and granular and often change into the large, segmented, keeled scales of the tail. The large, smooth ventral scales usually form eight longitudinal rows. Males may be immediately recognised by the preanal pores and large femoral pores.

Cnemidophorus exsanguis Lowe, 1956
Chihuahuan Spotted Whiptail

Distribution: Southeast Arizona and central New Mexico. In the south to Texas and Mexico.
Description: This species may reach a total length of 31.4 cm. They have six light stripes, bordered by brown bands, broken by light spots. There are five to eight rows of scales between the light stripes down the centre of the back. The underside is uniform light grey or white.
Habitat: Deserts, dry grassland, woodland, especially with a mixture of trees such as oak and pine.
Lifestyle, husbandry, and reproduction: These diurnal lizards hunt for insects and

spiders beneath leaves and fallen trees. Over short distances they can run on their hindlegs. These lizards are parthenogenic; small males are unknown. From June to August, females lay one to six eggs from which, after 6 months, the young hatch. In the vivarium they should be kept in the same way as *Cnemidophorus sexlineatus*.

Cnemidophorus gularis Baird & Girard, 1852
Texas Spotted Whiptail

Distribution: Southwest New Mexico to southern Oklahoma, in the south from Texas to New Mexico.
Description: These slender lizards may reach a total length of 20 to 27.9 cm. Along the sides there are seven or eight light stripes that are bordered by light-spotted, dark brown to green bands. Broader light stripes, sometimes divided into two, run along the centre of the back. In some populations, the stripes may be absent or only very poorly defined. Males have a red throat, light blue on the sides of the belly, and usually black spots on the chest. Females have a pink throat and white belly. The tail is pink to a brownish-orange. Juveniles have light stripes and a reddish tail.
Habitat: Semi-arid prairies and grassland, open land with bushes. Also in dry river beds and canyons.
Lifestyle, husbandry, and reproduction: These diurnally active lizards search in the sand and beneath rocks for food. Although in the wild they feed mainly on termites and grasshoppers, they will eat other insects and spiders in captivity.

Mating takes place in April and May. Between May and June, females lay from one to five eggs. A second clutch usually follows at the end of July. They may be kept in the same way as *Cnemidophorus sexlineatus* but in more spacious vivaria.

Cnemidophorus inornatus Baird, 1858
Little Striped Whiptail

Distribution: New Mexico, especially in southern Arizona. Also in western Texas and in the south towards Mexico.
Distribution: Only very few specimens reach the maximum total length of 24 cm. On the unflecked upper side there are six to eight dark reddish-brown to black stripes. Throat and belly are blue. Males are more brightly coloured than females. At its base, the tail is brown changing to blue further down.
Habitat: Dry to slightly moist grassland with some bushes.
Lifestyle, husbandry, and reproduction: In the wild, these lizards feed on beetles, grasshoppers, and spiders which are its main diet. When in danger they conceal themselves beneath rocks or amongst piles of wood. For their living quarters they often use the burrows of small mammals. They are usually found on land with sparse vegetation.

Mating takes place in spring, females lay two to four eggs between May and July.

These eggs hatch after some 6 to 8 weeks. This species should be kept in the same way as *Cnemidophorus sexlineatus*.

Cnemidophorus sexlineatus (Linnaeus, 1758)
Six-Lined Racerunner

Distribution: Florida westwards to Texas, New Mexico, Colorado, southeast Wyoming, South Dakota.
Description: These lizards seldom reach the maximum length of 26.7 cm. Along the back there are six or seven light stripes that are separated by dark green to black bands. Occasionally there is also a narrow yellow stripe along the centre of the back. The dorsal scales are smooth and granular. In males, the throat is green or blue. Females have a white throat. Their belly is also white or blue. The brown tail has stripes along the sides. Juveniles have a light blue tail.
Habitat: Dry sunny areas, open woodland, usually on dry ground.
Lifestyle, husbandry, and reproduction: These lizards are mainly active in the morning when they search for insects and spiders. During cold weather and at night they burrow underground.

In the wild, mating takes place from April to June. In June or July, females lay from one to six eggs. A second clutch follows around 3 weeks later. Newly hatched young appear from June to September. These lizards should be kept in a dry vivarium with some moist places and several hiding places. Daytime temperatures of 28 to 40 °C are needed, cooler at night. The vivarium should be sprayed each morning so that the lizards may lick up the resulting water droplets. A spotlight should be installed in the vivarium to produce a localised temperature of around 40 °C. The main diet of this species is insects, especially field and house crickets. It is important that vitamin and mineral supplements are given regularly. A 2-month hibernation at around 10 °C will bring these lizards into breeding condition. Females are gravid for 3 to 4 weeks before laying up to six eggs which they bury in the sand. The eggs should be placed in an incubator and be kept at 25 to 30 °C. The young hatch after 2 to 2 1/2 months.

Cnemidophorus tigris Baird & Girard, 1859
Western Whiptail

Distribution: Northern Oregon, southwest Idaho to lower California and Mexico. In the east to western Texas.
Description: The Western Whiptail can reach a total length of around 30 cm of which a little more than half is the tail. Its slender body is barely broader than the index finger and is very streamlined. The legs and toes appear very long in comparison to the body. On the light grey to beige background, there are often four to eight lighter stripes. Frequently, many dark lines and flecks are also found. Throat

and belly are usually white or yellow, very rarely also completlely black. There are often black spots on the chest. The grey or greenish grey tail normally has black speckling on the sides. Juveniles are light blue.
Habitat: Sandy deserts with sparse vegetation, bushy hillsides and mountain slopes with light woodland, from 0 to 1500 m above sea level.
Lifestyle, husbandry, and reproduction: These excitable lizards live in deserted small mammal burrows, and when disturbed run rapidly to safety. A well lighted vivarium with a ground area of at least 150 × 100 cm is needed for these lizards. A suitable substrate is medium grade sand over which loam is scattered. A tree root or flat, stacked rocks will provide adequate refuges. A spotlight should provide a localised temperature of 40 to 50 °C. Ultraviolet light should also be given at regular intervals. Good ventilation is essential. At night the temperature should fall to around 18 to 20 °C.

To cover their liquid requirement it is sufficient to give them a shallow bowl of drinking water twice weekly. This should be removed after the lizards have slaked their thirst. In the wild, Running Lizards eat termites, beetles, butterflies, crickets, grasshoppers, and flies. A similarly varied diet should be given in captivity.

During the summer months it is expedient to keep these lizards in an outdoor enclosure. For breeding, Switak (1982) advises a cool hibernation at 10 to 12 °C from late November until late February, before which the lizards must be fasted for around 2 weeks. The hibernation containers must be completely dry.

Dracaena Daudin, 1802
Caiman Lizard

Two species of this genus live in tropical South America, mainly along the banks of the Selva. They can grow to well over 1 m in length with a stocky body and powerful tail that is compressed at the sides. The dorsal scalation consists of large keeled plates surrounded by smaller granular scales. On the upper side of the tail there are two longitudinal keels, similar to those of crocodiles.

Dracaena guianensis (Lacépéde, 1788)
Guyana Caiman Lizard

Distribution: Northeastern South America.
Description: As described for the genus. Caiman Lizards reach a maximum length of 120 cm and are dusky olive to dark brown in colour.
Habitat: Swampy areas with rich vegetation.
Lifestyle, husbandry, and reproduction: Caiman Lizards are mainly ground dwellers but they will climb into riverside bushes to sleep in relative safety. These lizards feed mainly on soft aquatic animals such as water snails, the shells of which they are easily able to break. Their eggs are often laid in termite mounds.

Dracaena guianensis

To house a pair of Caiman Lizards, a vivarium of at least 200 × 200 × 150 cm is needed, the base of which should be covered by a layer of washed gravel and a further 10 cm layer of sand or peat. One-third of the base should be a water section at least 15 cm deep. Tree roots or stout branches will allow the lizards to climb at will.

Häberle (1973, 1979) obtained a Caiman Lizard that was in extremely poor condition. It was obvious that it had not eaten for several weeks. Because in the wild this lizard feeds mainly on water snails and other aquatic molluscs, Häberle chopped up two hairless rat pups which he mixed with the flesh of two large edible snails. The lizard tended to select mainly snail flesh but small amounts of rat flesh were also eaten. After some weeks the lizard became accustomed to eating unadulterated rat flesh. Before being given to the lizard, the flesh of the snails and rat pups was injected with a liquid multivitamin preparation.

Lizards 237

Pholidobolus montium

Pholidobolus Peters, 1862
Dwarf Tegu

The two species of this genus live in the Andes of Ecuador and Peru. They reach a total length of 15 cm, of which around two-thirds is the tail.

Pholidobolus montium (Peters, 1862)
Dwarf Tegu, Montane Dwarf Tegu

Distribution: Ecuador, around the Quito region up to an altitude of 2,850 m, in the Tierra Fria (Cold Land).
Description: The ground colour of these Tegus is greyish brown. Juveniles have a brown back. With increasing age they become more grey with an insignificant nar-

row black central stripe. Along the sides there are two white longitudinal stripes that stretch from the snout to the root of the tail. Sometimes there are two brown-black stripes running down the white underside which is also occasionally spotted. Frequently, between the two white stripes there is an additional narrow white stripe which starts behind the forelegs.
Habitat: Sandy, rocky ground with sparse vegetation and low-growing succulents. Also on grassland with some bushes.
Lifestyle, husbandry, and reproduction: Dwarf Tegus require a spacious vivarium. Brauer kept two males and three females in an 80 × 22 × 23 cm vivarium containing a 5 cm layer of fine sand as substrate. The furnishings were root stocks, pieces of bark, rocks, and bunches of dry grass. These lizards require a daytime temperature between 20 and 25 °C and will feed on all manner of arthropods.

Before copulation takes place the male bites the female firmly on the neck. The two 1-cm-long eggs are buried in the earth and may take up to 216 days to hatch (Bauer, 1980).

Tupinambis Daudin, 1803
Tegu

Tegus occur over an enormous distribution range which stretches from South America, east of the Andes to northern Argentina. Here, these powerful lizards live in forests and become partial followers of civilisation. They may reach a total length of 140 cm. Tegus have a pointed head and can withdraw the tongue into a sheath. The tail is double the SVL. The ventral scales are significantly larger than the small granular dorsal scales. However, in comparison with other Teiids, these scales still appear small. They are arranged in 22 to 27 longitudinal rows. Two species are known.

Tupinambis teguixin (Linnaeus, 1758)
Black and White Tegu

Distribution: From tropical and subtropical Columbia, Venezuela, and Guyana through the Amazon Basin of Venezuela and Columbia along the larger rivers to Paraguay and northern Argentina.
Description: This Tegu can reach a maximum length of 140 cm and has 17 to 19 preanal and femoral pores. The ground colour is blackish brown with a light blue sheen. Along the back there are numerous transverse bands composed of yellow spots of various sizes. These bands extend to the tip of the tail. There is a series of irregular whitish yellow flecks on the head, throat, and limbs.
Habitat: Tropical rain forests, savannahs and flatlands with thorn bushes. Along rivers in hilly grasslands with subtropical trees. Also found on coastal, sandy areas with grass and trees (Köhler, 1989).
Lifestyle, husbandry, and reproduction: As ground dwellers, these Tegus are often

Tupinambis teguixin

found in forest clearings. They dig their own burrows and are excellent climbers and swimmers. In the wild, their diet consists of arthropods, small vertebrates, eggs, and occasional vegetable material such as fruit, flowers, or herbs. Females often lay their eggs in termite mounds where the eggs are kept at a constant temperature, and humidity is favourable to their development. After hatching, the young break out from the termite mound.

Because of their size, these burrowing lizards require a very spacious vivarium with a deep, sandy substrate which should be kept constantly moist. Hiding places and refuges are vital and may consist of stacked, flat rocks, firmly cemented together, hollow tree trunks, or root stocks. Stout climbing branches are also essential. A large water bowl will allow the lizards to drink and bathe at will.

An adult will require a vivarium of at least $200 \times 100 \times 100$ cm. Köhler (1989) kept his Tegus in a $220 \times 160 \times 200$ cm vivarium containing a substrate of newspaper, plastic grass, or wood shavings. Köhler does not consider a natural substrate to be essential and prefers the synthetic substitute for reasons of hygiene. An important part of the vivarium was the water container in which the lizards bathed daily.

As refuge, a piece of artificial leather 80 × 60 cm was used. This was weighted down with rocks at the corners, but the lizards were still able to crawl beneath it. One-quarter of the vivarium base was warmed by a heating cable, whilst a spotlight provided a localised temperature of 40 to 45 °C, a place which the lizards frequently used for basking.

According to season, the temperature was set at various levels; in the morning 20 to 25 °C and in the afternoon 22.5 to 27 °C. In the morning the humidity was around 70 to 100%, falling to around 55 to 80% in the afternoon. Throughout the year, the lights were on for 12 hours daily. Each afternoon an Osram Ultravitalux™ lamp, installed at a height of 1 m, was switched on for 20 minutes. Twice weekly the Tegus were given mice, Argentinian cockroaches, fish, earthworms, locusts, and crickets as well as banana, grapes, cherries, and sprouting soya beans, sunflower seeds, and lentils.

The male came into breeding condition during April, May, and August, but copulation was not seen until November. The male followed the female, hissing loudly and gently biting her tail. When the female was ready to mate, this was indicated by raising the tail and energetic head shaking. Köhler saw that before extending the hemipenes, the male rubbed the root of his tail along the ground, then massaged his cloacal area with his hindlegs. Before the actual copulation. the male holds the female firmly by a bite to the neck. In captivity, females lay eggs throughout the year. In Köhler's case the eggs were always laid beneath the artificial leather.

A clutch may consist of 4 to 32 eggs measuring 42.0 to 54.0 × 25.5 to 31.0 mm and weighing between 17 and 24 g. The eggs have a firm, elastic shell. Placed in slightly moist sand and at a temperature of 30 °C and humidity between 80 and 100%, the eggs hatch within 152 to 171 days. On hatching the young measure between 22.2 and 24.7 cm. They have a SVL of 8.0 to 8.7 cm and weigh approximately 18 to 20 g. During incubation the eggs increase in girth by around 25% and weight by 55%. Only 2 days after hatching, Köhler's young ate raw egg. The hatchlings may be reared on nestling mice, insects, and fruit flies to which regular vitamin and mineral supplements are added. After only 1 year the young can reach a length of 40 cm.

In Langerwerf's case (written communication), in Alabama (United States), four Argentinian specimens lived in a 3 × 3 m outdoor enclosure with a very deep substrate. Although in the Cordoba area regular night frost may be expected from May to September and temperatures of -5 °C are not uncommon, the lizards were kept in this enclosure throughout the year. Alabama also has frost and snow in winter. At the end of August, Langerwerf's Tegus were no longer regularly fed although the temperature was still around 30 °C during the day. At this point the lizards began to dig tunnels some 50 to 70 cm deep into which they retired from mid-September until mid-March (6 months!). Langerwerf's Tegus never deviated from this annual cycle. In early June, one female tried to lay her eggs as far away from her companions as possible. She broke out of the enclosure and laid the eggs in a corner of the adjacent garden, where afterwards she enjoyed a good meal. Langerwerf's

Tegus were fed on rats, chicks, mealworms, and eggs. In early October, Langerwerf discovered seven newly hatched young in the enclosure that he had placed around the eggs in the garden. The young had a SVL of 8 cm and a tail length of 12 cm. Even at very low temperatures they proved to be quite voracious feeders. At the time the young hatched, the parents were already in hibernation. Incubation took only three months. Langerwerf suspects that in such an enormous distribution range, there must be subspecies.

WORM LIZARDS
Suborder Amphisbaenia

The systematic placement of the Amphisbaenians was, and still is, the subject of great debate amongst herpetologists, even to the extent that they should be assigned to the order of Squamata because "... this is proved by the structure of integuments, egg tooth, plagiotreme cloaca and paired hemipenes of the males" (Böhme, 1988). In original forms *(Bipes, Blanus)* the hemipenis is undivided, whilst in further evolved species it is deeply cleft *(Trogonophis)*.

Amphisbaenians live in the tropical and temperate regions of the New and Old Worlds. There are around 20 known genera divided into four families: *Bipedidae, Amphisbaenidae, Trogonophidae*, and *Rhineuridae* (Gans, 1978).

According to species, Amphisbaenids reach very different lengths. The smallest are only 10 cm long whilst the largest may exceed 80 cm. They all have a cylindrical body and an extremely short tail. The head is proportionately large. The lower jaw appears to be almost sunken into the upper jaw so that the narrow mouth is always overshot. The teeth of the upper jaw fit in between those of the lower jaw. The remnants of eyes may still be seen as small dots beneath the skin. The most Amphisbaenids can distinguish with these eyes is the difference between degrees of light. The ears are also covered by skin, but are still excellent at detecting ground vibrations.

Whilst in all other worm-shaped lizards or snakes the lungs are double-developed or only the right lobe of the lung is underdeveloped, in Amphisbaenids it is only the right lobe of the lung which is functional. The left lobe is only rudimentary. Only the genus *Bipes* has forelegs; in all other species the limbs are completely absent.

Frequently, typical lizard scalation is only present in the head region. The other skin is arranged in more-or-less complete rings, whereby two body rings form a vertebra!

In accordance with their subterranean lifestyle, Amphisbaenids are mainly red, brown, or grey in colour. These animals rarely come to the surface, even at night. However, they are occasionally found beneath rocks, especially in dry areas.

Amphisbaenids build an extensive system of underground tunnels around which they can move very quickly, both forwards and backwards. Their prey consists mainly of insect larvae and earthworms, the presence of which they detect from vibrations in the ground. Some species live in ant or termite nests. Here they find not only sufficient food, but also an excellent place to lay their eggs. To drink, Worm Lizards suck up water with the lips only slightly above the ground. Most species lay eggs, the size of which may vary according to the number in the clutch. Only very few species give birth to live young.

Because of their secretive, underground lifestyle, Amphisbaenians are not particularly popular amongst herpetologists, although many of them are extremely long-lived. The conditions for keeping the various species of Worm Lizards correctly are generally very similar.

They require a vivarium with a very deep layer of substrate. Even for species from dry areas, part of the substrate should be kept constantly moist, since it is here that the animals will satisfy their liquid requirement. For nontropical species, a mixture of loam and clay is a suitable substrate. This will enable the lizards to construct tunnels that will not collapse.

According to their origins, Amphisbaenids require a daytime temperature of 20 to 28 °C, falling by a few degrees at night. Species from temperate regions are hardier and easier to breed if they are given a seasonal cycle including 4 to 6 weeks of hibernation. Occasionally the substrate must be changed depending upon the extent of fouling. In a small vivarium or one used for larger species, the substrate should be changed every 3 to 6 months.

An all-glass aquarium is especially suitable for housing Amphisbaenids. If the glass panes are darkened, the movements of the animals in their tunnels may be observed. Suitable means of darkening the glass panels are pieces of cardboard or plywood.

True Worm Lizards
Family Amphisbaenidae

The representatives of the *Amphisbaenidae* live predominantly in tropical forests, although some do inhabit drier regions. This populous family has been divided into two subfamilies. *Amphisbaenidae* differ from *Trygonphidae* in that their teeth are on the side edges of the jaws (pleurodentes bite) and their tail is less pointed and does not bend downwards. Very little is known about most genera and species.

Blanus Wagler, 1830
Ringed Worm Lizards

The *Blanus* species live in Africa, north of the Sahara Desert, in southwest Europe and western Asia on dry hillsides, edges of light woodlands, the edges of lakes, and in meadows. In this species, one ring of skin corresponds to one vertebra. In other species of Worm Lizards, two rings comprise one vertebra.

Blanus cinerus (Vandelli, 1797)
Moorish Worm Lizard, Ringed Worm Lizard

<u>Distribution</u>: Central and southern Iberian Peninsula and Morocco. In Spain up to an altitude of 1,400 m.
<u>Description</u>: In build and colour, the Moorish Worm Lizard is very reminiscent of a large earthworm. Its small pointed head with vestigial eyes is barely distinguishable from its tail. The scalation on the body forms equal, transverse rings, whereby the surface appears to have a gridlike pattern. On the body there are 110 to 125 rings and on the tail 20 to 22. There is a longitudinal fold of skin along each side of the body. This Worm Lizard can reach a length of up to 30 cm. These lizards are pale pink in colour but there are specimens that are a darker brownish red.
<u>Habitat</u>: This species does not like moisture and prefers dry, loose ground in pine forests or cultivated land.
<u>Lifestyle, husbandry, and reproduction</u>: Because the Moorish Worm Lizard avoids daylight and lives almost exclusively underground, they are usually only found by accident, mainly in spring when rocks are turned in the early hours of the morning. These lizards return to their burrows in late morning and only reappear above ground during the night. This has also been proved by the observation of captive animals. In southern Spain (Cadiz), Busack (1978) occasionally saw Worm Lizards active on the surface during the day.

When examined, the contents of the stomach of a *Blanus* included the remains of small worms, ants, ant larvae, caterpillars, beetle larvae, and other insects. If there are termite hills in the vicinity, Worm Lizards feed almost exclusively on these insects. As well as insects, Worm Lizards will also eat spiders. In captivity, these Worm Lizards may be fed on small ground-dwelling insects, small earthworms, and small mealworms.

The Moorish Worm Lizard is oviparous. In July, a 170 to 180 mm female from Morocco laid a single egg measuring 27×5.2 mm. Further information on reproduction, i.e., duration of gravidity etc. is unfortunately unknown. In this respect, vivarium study of these animals could yield a wealth of new knowledge.

Kenneweg (1956) obtained a single specimen from Portugal, which he placed in a $20 \times 15 \times 18$ cm aquarium containing a 6 cm substrate of moist sand and loose garden loam upon which a number of small pieces of sandstone and clumps of moss were placed. By the following morning there were already a number of tunnels in which the lizard's activities could be seen. The lizard was at first fed on small centipedes and ants, after which mealworms were added to its diet. On warm summer nights the lizard was often seen on the surface of the substrate. As soon as the light

Blanus strauchi

was switched on however, it quickly disappeared into its burrow. The vivarium was occasionally sprayed after which the lizard would push its head out from one of the holes to suck up the resulting water droplets from the glass sides of the aquarium.

Blanus strauchi (Bedriaga, 1884)
Turkish Worm Lizard

Distribution: In most parts of Turkey and on the Greek islands of Kos and Rhodos.
Description: This Worm Lizard is similar to the western European species and reaches a length of 30 cm. The body has 110 to 112 rings, the tail 18 to 20.
Habitat: See *Blanus cinerus*.
Lifestyle, husbandry, and reproduction: Because of the heavily boned head, this species can burrow through the earth very quickly. When moving through their burrows, one may see wavelike movements running along the body.

In spring, these lizards can sometimes be found in pairs beneath rocks. Their food, e.g., small insects or ground beetles, can be detected aurally by the ground vibrations they cause. Once they have caught their prey they hold it firmly, tearing-off pieces by rotating the body. Their liquid requirement is satisfied either from their food or from moisture in the ground. In inclement weather, times of drought, or in winter, these Worm Lizards burrow deep into the ground. It is thought that Turkish Worm Lizards also lay eggs. According to Sauer (1989) these lizards are sexually mature at 3 years of age.

Florida Worm Lizards
Family Rhineuridae

Rhineura Cope, 1861

The sole recent representative of this genus is *Rhineura floridana*. The head of this slender creature is very horny and the snout somewhat flattened and shovel-like. The lower jaw is slightly undershot. The tail is also slightly flattened.

Rhineura floridana (Baird, 1858)
Florida Worm Lizard

Distribution: North and central Florida.
Description: These Amphisbaenids usually reach a length of around 30 cm; only very rarely do they reach the maximum length of 40 cm. They are reddish brown to pink in colour, and despite their rough skin they are very reminiscent of earthworms.
Habitat: Dry, sandy areas.
Lifestyle, husbandry, and reproduction: The Florida Worm Lizard lives exclusively underground and normally only comes to the surface after heavy rainfall. Spiders, earthworms, and termites form the main part of their diet. In summer, females lay one to three eggs from which the young hatch in early autumn. Because of their subterranean lifestyle, not much is known about these animals, even in captivity.

Sharp-Tailed Worm Lizards
Family Trogonophidae

Sharp-Tailed Worm Lizards live in northwest Africa, Somalia, Ethiopia, on the island of Sokotra, in Arabia to southern Iran. This is a very original Amphisbaenian of which four genera and a total of six species are known. They all live in dry areas and with a maxium length of 20 cm they are some of the smallest Amphisbaenians.

The body is compressed and the head short with a rounded snout. The tail is pointed and bends downwards. Some species are ovoviviparous whilst others lay eggs.

Trogonophis Kaup, 1930
Checkerboard Worm Lizard

Only one species; in northwest Africa in diverse habitats.

Trogonophis wiegmanni Kaup, 1830
Checkerboard Worm Lizard, Double-Ended Worm Lizard

Distribution: Morocco to Tunisia.
Description: This species reaches a length of 20 cm. According to where they are found, they are a conspicuous bright yellow with dark brown to black flecks.
Habitat: Open steppes, mountainsides, and forested areas, mainly in sandy areas near the coast; never found in completely dry areas.
Lifestyle, husbandry, and reproduction: These lizards move in an "accordion-like" fashion and come to the surface more frequently than other Amphisbaenians, sometimes even to bask. However, the head and hindquarters are always kept below ground. They enter and leave their burrows depending upon the amount of moisture in the ground, but at night they come to the surface. During summer and winter they remain underground. Obviously ants and termites are their main food. Their preferred temperature is around 29 °C. Females are ovoviviparous, giving birth to one or two quite large young. These Worm Lizards are sexually mature at 2 1/2 years old, but females do not produce young every year. Young are usually born in September.

These Worm Lizards require a vivarium of 50 × 70 cm. A suitable substrate is a 10 to 15 cm layer of sand and loam upon which several flat rocks are placed. As well as ants, in captivity these Worm Lizards will also eat other small insects which should be liberally coated with calcium and a powdered vitamin supplement. The breeding condition will be stimulated if the lizards are allowed to hibernate for 4 to 6 weeks at around 10 °C.

Tuataras
(Order Rhynchocephalia)

Bridged Lizards are representatives of the order Rhynchocephalia, which during the Triassic and Jurassic Ages were quite common. The only survivor from these times is the Tuatara which belongs to the family Sphenodontidae and the genus *Sphenodon*. There are several subspecies.

Because of the complete protection afforded to these animals by the government of New Zealand, they will never be seen in a private collection.

Sphenodon punctatus (Gray, 1842)
Tuatara

Distribution: Only on around 30 islets in the Cook Straits and between the Bay of Plenty and the Bay of Islands to the west of New Zealand.
Description: In build and appearance, the Tuatara closely resembles an agama. The pupils are vertical and there are no ear orifices. They reach a length of 65 to 74 cm. Males are larger than females. The name "Bridged Lizard" refers to the complete, double bridge of bone over the temporal area. This is an anatomical peculiarity which in present-day reptiles is only found in crocodiles. Their skin is more reminiscent of a toad than a reptile. The spines and warts are soft to the touch. These very rare creatures have a so-called vestigial "third eye" which is only very rudimentary and which is also found in other species of lizards, but in the Tuatara it is more original. Only the iris is absent. It is connected to the brain by a nerve. The "third eye" is thus not an organ of vision and only serves to detect light and possibly regulate the body temperature.

In Tuataras the teeth are at the edges of the jaws and with increasing age become worn away to a knobbly stump as is the case with agamas and chameleons. New teeth do not grow, and thus with the passage of time, the teeth are eventually completely worn down. In old age they bite only with the edges of the jaws.

Three subspecies are known: *Sphenodon punctatus punctatus*, *Sphenodon p. guentheri*, and *Sphenodon p. reischeki* which some herpetologists consider to be a separate species.
Habitat: During the day these lizards live in the burrows of sea birds.
Lifestyle, husbandry, and reproduction: Despite the very strict conservation and

Sphenodon punctatus

protection laws of the New Zealand government, the numbers of these unique lizards are still threatened by animals that have been introduced by humans, i.e., cats, dogs, pigs, goats, rats, and mice. Rats are thought to catch hatchling Tuataras. Moreover, the reproductive cycle of the Tuatara is very unpredictable. These animals are only sexually mature at an age of around 20 years. Because of very strict protection and conservation measures imposed by the New Zealand government, the constant monitoring of the lizards and the removal of introduced predators, the number of Tuataras has happily increased over the last decade.

In the wild, the Tuatara mates from February to March. As opposed to females, males are capable of breeding every year. During the breeding season, males claim a territory and begin courtship activity. A territory ranges over an area of 25 m^2, at the centre of which are the burrows where the male lives. Males emerge at dusk and sit at the entrance to their burrow throughout the night, when in appearance and behaviour, they are much more aggressive than usual, with erect neck comb. Any male straying into the territory will immediately be attacked. Older males often have scars from earlier bite wounds. The number of males and females in any given population is approximately equal, hence the number of males willing to mate far exceeds the number of females in the same condition. Courtship begins when a female enters a male's territory. Raising the body on stiffened legs, the male circles the female. Males do not have an elongated genital organ. They transfer sperm by pressing the cloaca against that of the female. Copulation may last for up to 1 hour.

In November, females lay eggs from which the young hatch around 1 year later. Usually only around one-half or one-third of the eggs develop. The others die off or fall victim to mould. Females are gravid only every 4 years.

On the island of "The Brothers," the eggs are collected annually and are artificially incubated at the Victoria University of Wellington at fluctuating temperatures between 18 and 23 °C. Constant temperatures reduce the incubation time, so that the young generally hatch in May. A clutch usually contains 5 to 15 eggs. Females often go up to several hundred metres from their burrows to find a suitable nesting place where they may spend several days or even weeks. Here they dig a shallow depression in which the eggs are laid. When an egg is laid it is carefully covered. The female returns to the nest each day for around 1 week to protect the clutch and to prevent another female from using the same nest site. In the wild, the 10 cm-5 g young only hatch after 12 to 15 months.

In contrast to the adults, the young are active during the day, thus preventing them from being eaten by adults. They only join the colonies of adult Tuataras at the age of around 1 year. The poor breeding rate of the Tuatara is only equalled by its longevity. According to Henle (1981), these lizards may live for up to 150 years. A life span of at least 77 years has been verified.

In the areas where colonies congregate, there are burrows where the Tuataras live and which they share with sea birds of the genus *Puffinus*. This cohabitation is usually quite harmonious and advantageous for the Tuataras. The excrement from the birds attracts insects upon which the lizards feed, mainly beetles, large wingless grasshoppers, earthworms, and snails. They will however, also occasionally devour nestling birds and birds' eggs. Large Tuataras will sometimes eat older and weaker specimens of the smaller sea bird species.

As opposed to other reptiles, the Tuatara prefers temperatures between 17 and 20 °C, and are still quite active at lower temperatures. This is why they are active at dusk and during the night. The Tuatara does not hibernate in winter. Instead its metabolic rate and activities are reduced.

In 1952, the government of New Zealand presented individual specimens to various zoos (London, New York, San Diego) but none of them survived. At that time it was not known that the Tuatara prefers lower temperatures, hence the temperatures at which they were kept were too high. Later, Basel Zoo (Switzerland) received a Tuatara (Stemmler-Morath, 1958) which, when placed in its vivarium, hid beneath a large bunch of moss. When disturbed it did not bite. It almost always defecated in the water bowl and only very rarely on land. It was fed on Wall Lizards, Capricorn Beetles, crickets, and mealworms, but refused earthworms, snails, nestling mice, rats, hamsters, meat, and fish. After 1 month, the lizard laid two white soft-shelled eggs at the most moist place in the vivarium, beneath the moss. The eggs were 21 × 13 mm and weighed 2.5 g. Later, a further egg was produced but unfortunately none of the eggs developed. Soon afterwards the lizard died as a result of a liver infection caused by parasite poisoning.

Frank & Bachmann (1976) reported on a further Tuatara that was kept in a large vivarium with a ground area of 130 × 130 cm. The centrepiece of this vivarium was a fibreglass mound with several preshaped burrows. The Tuatara could choose between several temperature ranges: 12 °C in the burrows, outside the burrows 14 to 16 °C, on the sides of the mound 18 to 20 °C, and at the top of the mound 20 to 22 °C. This animal was fed on large insects, earthworms, young mice and rats. It died after eight years as a result of an injection.

In 1908, the Zoological Institute of Upsala (Sweden) obtained two Tuataras, one of which soon died. According to Cyren (1934) the other lizard lived for 23 years in a wooden crate measuring 75 × 40 × 40 cm, containing only wood-wool and a water container 7 cm high and 30 cm in diameter. The crate was covered with wire mesh. The crate always stood on the floor beneath a workbench and, from 1916 onwards, in a damp cellar with a winter temperature of 16 to 18 °C, rising in summer to 20 °C.

In summer, the Tuatara was placed in a wooden crate of 190 × 75 × 30 cm that had been divided into three sections. The first contained only wood-wool, the second a large water container, and the third large sods of lawn turf. Within only 1 week the Tuatara began to dig burrows in which it remained during the day. As soon as the cardboard cover was placed over the third section, giving the impression of a burrow, the Tuatara often spent the day lying on the top of the turf and making no attempt to hide. It was fed once each week and was given 15 to 20 earthworms. During the following week it was given strips of raw meat.

After being placed in the larger container, the Tuatara was much more active. It bathed frequently, enjoyed swimming, and often lay in the water for long periods. It sloughed only once each year, between January and February. Occasionally, in the early hours of the morning, and irrespective of the time of year, the Tuatara would emit a short, repeated, single-toned squeak.

A pair of Tuataras kept at Chester Zoo (United Kingdom) and a male at the Senckenberg Museum in Frankfurt (Germany) also lived for a considerable length of time. The animal in Germany was presented to the renowned German herpetologist Robert Mertens as a gift from the New Zealand government in recognition of his distinguished service to herpetology. Mertens unfortunately died in 1975.

The Zoological Gardens of Berlin received 10 Tuataras, approximately 4 years old, in 1990, as a gift from Wellington University (New Zealand). In Berlin, these lizards live in a 25 m^2 vivarium that is air-conditioned at around 16 °C and that has an artificial "stream" flowing through it. There is a separate, preformed burrow for each Tuatara.

Crocodilians
(Order Crocodylia)

Crocodilians live in the subtropics and tropics of the Old and New Worlds and are more closely related to birds than to any other reptiles living today. There are now only three surviving families: *Alligatoridae, Crocodylidae*, and *Gavialidae*.

Their lizardlike build and more-or-less "bony" skin leads to the common name "Armoured Lizards." These species can grow to lengths of barely 140 cm to almost 10 metres.

Crocodiles are perfectly adapted to their amphibious lifestyle. Even when swimming underwater, the eyes, ears, and nose can be above the surface. They can also breathe whilst their body is underwater by opening the mouth. This is made possible by a piece of skin in the pharynx that separates the air passages from the oral cavity. Moveable eyelids and a semitransparent nictating membrane can cover the eyes. Further folds of skin seal the nostrils and external ear orifices. The head, and especially the snout, is developed differently in each species. There are species with a very long snout and others with a very short snout. Likewise there are those with a broad snout and others with a very narrow snout.

Because of their enormous lungs, these animals are able to slow down their metabolic rate and may remain submerged for more than an hour. All Crocodilians are excellent divers and swimmers. The forelegs have five toes whilst the hindlegs have only four. There are claws on only the first three fingers and on all four toes. The hindfeet are webbed and are a little longer then the forefeet. Although rather clumsy on land, crocodiles are capable of an amazing turn of speed. In the water, the long, muscular tail acts as a propeller when swimming or diving.

The skin on the back of Crocodilians is especially hard and horny, with many bony plates. The upper side of the tail has a row of upright scales along each side. These rows of scales join together around the centre of the tail, and continue as a comb to the tip. The arrangement of the keels on the tail serve to distinguish the various species; other means of distinguishing species are the number of scales and plates as well as the position of the rear occipital and nuchal scales.

Crocodilians are constantly growing new teeth. Old teeth fall out after a while to be replaced by new ones. The cone-shaped teeth have a long root, open at the bottom, and fit into alveoles. The various species of Crocodilians may also be distinguished by the number of teeth.

All Crocodilians feed on other animals. Those with the longest snout have evolved to become specialist fish eaters. Most crocodiles devour their prey in water, but they can eat on land. If the prey or piece of food is too large, the crocodiles hold it in the mouth and use their limbs to tear off edible-sized pieces. To do this they rotate on their own axis in the water. The amount of food required by Crocodilians is quite small in comparison with warm-blooded animals of a similar size. Although classed as ectothermic (cold-blooded), crocodiles are able to regulate their heat somewhat.

During the breeding season, pairs of musk glands on the chin and cloaca emit a secretion that has a stimulating effect upon sexual activity. Mating takes place in water. Males have an unpaired, protrudable penis. The cloacal slit runs in the same direction as the body axis. The hard-shelled eggs have a rough surface. According to age and species, females may lay from 15 to 100 eggs, each of which is 7 to 9 cm long. At temperatures of 29 to 34 °C and almost 100% humidity, incubation takes between 90 and 115 days. Females actively tend their offspring. Not only do they build and guard the nest, but after hatching they carry the young to the water in their mouth. Two types of nests are built: the first dug into sandy ground and the second consisting of a pile of decomposing vegetable matter. Once maternal obligations are completed, the hatchlings live away from the parents since cannibalism is usual amongst crocodiles. On crocodile farms, only animals of the same size are ever kept together in groups.

In earlier times crocodiles were relentlessly hunted. Nowadays, habitat destruction is the main threat to these animals. Crocodilians do, however, have an important ecological function, such that, should they become extinct, there would be severe consequences for mankind. In South America the decimation of caiman allows piranhas to breed and multiply at an unnatural rate. Piranhas eat all manner of fresh-water fish which for local inhabitants is an essential source of food. Similar results have been experienced in Africa and Australia. The decline in the numbers of freshwater food fish—when considered in connection with caiman—may be partly caused by the fact that the excrement from these lizards, which regularly fertilised the waters and provided food for lower organisms, is no longer present. Thus food for young fish is no longer available.

Because almost all species of crocodiles are now severely threatened, they should only be kept and bred by farms and zoological gardens. Hatchlings also grow so quickly that after 2 or 3 years, they can no longer be kept correctly in a private collection. Zoos also rarely have space for these unwanted animals which have outgrown their quarters in a private collection. Nowadays, zoos no longer have vast collections of individual species, preferring to keep larger breeding groups of fewer species.

If housed and cared for correctly, crocodiles are very long-lived. Spacious enclosures are the main prerequisites for the correct husbandry of these animals. When crocodiles are kept, every attempt should be made to breed them. This is not entirely true in the case of the Nile Crocodile; the Cologne Aquarium (Germany) has bred so many young of this species that they will no longer accept any new animals from any source (Jess, written communication).

When keeping Crocodilians, the following factors are vital:

- The land part of the enclosure must be large enough to allow the animals to build a nest.
- The water must be deep enough to allow the animals to mate.
- Any furnishings must be firmly anchored and easy to clean so that the animals can not injure themselves and the enclosure does not become excessively fouled.
- The water section must be connected to an overflow which enables excess food and excrement to be easily removed.
- Air and water temperatures should be regulated according to the origins of the animals. In general an air temperature of 27 to 30 °C and a water temperature of 25 °C are sufficient.
- As well as regular cleaning, a varied diet should be given in limited amounts since, in captivity, these animals quickly become obese or fall victim to rachitic deformities.
- To prevent deficiencies, calcium and vitamins should always be added to meat and fish.
- Inanimate food should be thrown to crocodiles or be given from very long forceps or the end of a long rod.
- Even so-called tame specimens should be treated with care and respect. They may still suddenly bite and cause severe injuries.
- The enclosure must be secure and allow no means of escape.
- It is strongly advised that the keeper be insured against bites and water pollution.

ALLIGATORS AND CAIMANS
Family Alligatoridae

The actual home of the the Alligatoridae is America, although one species lives in eastern Asia. There are four genera with a total of seven species, each of which has a differently shaped snout. That of *Alligator mississippiensis* is broad, flat, and long, whilst that of *Caiman latirostris* is extremely short. When the mouth is closed, only the teeth of the upper jaw can be seen. The teeth on the lower jaw fit inside the teeth on the upper jaw. Only the fourth tooth on the lower jaw fits into a corresponding depression on the upper jaw, but when the mouth is closed, this tooth is not visible. There are no comb-scales on the limbs. The tubercles on the neck are usually not widely separated from the tubercles on the back. The longest representatives of this family may reach a total length of over 6 m.

Alligators and Caiman feed on a wide variety of invertebrates, vertebrates, and carrion. Females lay their eggs in a nest of rotting vegetable matter and afterwards protect the nest.

Alligator Cuvier, 1807

Of the two species of Alligators, one lives in America and the other in China. Both have a flat, broad snout. They may be distinguished from Caiman by the bony, nasal septum and the six large, jointed nasal tubercles. Only the Chinese Alligator has ventral scales which are connected by bony joints, in which respect it is similar to the Caiman. Deraniyagala has suggested a separate generic name, "*Caigator*," for these animals. It is only in old age that some American Alligators develop relatively weak, bony connections between some of the ventral scales. In both species, the connections are in one part. In Caiman, the connections between the ventral scales always consist of two parts.

Alligator mississippiensis (Daudin, 1802)
American Alligator

Distribution: North America: North and South Carolina, Georgia, Florida, Alabama, Mississippi, Louisiana, Arkansas, southern Oklahoma, and southeast Texas.
Description: Although these animals may reach a length of up to 6 m, a more usual length is 3.5 to 4.0 m. The snout is flat, broad and rounded at the tip. On both sides of the jaw there are five teeth in the intermediate jaw. When the mouth is closed, the fourth tooth on the lower jaw is not visible; it fits into a groove in the upper jaw. The rear occipital scales are arranged in two or three transverse rows, behind which there are two pairs of larger and two pairs of smaller nuchal scales. The dorsal scales are arranged in 8 longitudinal rows and 17 or 18 transverse rows. The young are almost black with yellowish transverse bands. The light underside does not have any markings.
Habitat: Fresh and brackish water swamps, ponds, lakes, rivers, and their tributaries.
Lifestyle, husbandry, and reproduction: Alligators are only rarely available commercially although farms in the United States breed them regularly and in vast numbers. Their tail fat and lard is used in restaurants whilst their skin is in great demand in the leather industry.

In the wild, outside the mating season, the sexes live apart and in different areas. Males prefer large, open waters that are very deep. Females remain in smaller, shallower waters. In the breeding season, during April and May, males claim and defend a territory by loud roars and thrashing the head on the surface of the water. The largest, strongest males claim the largest territories. Courtship takes place over several weeks so that females usually mate with more than one male. American Alligators mate in shallow water. Before copulation they stimulate one another by rubbing heads together and loud coughing noises. When the female is receptive, the male mounts and mates from the side for several minutes.

In early June the female lays eggs which are then covered with a large mound of earth and decaying plant material. Nests may be up to 1 m high and 1.5 to 2.0 m wide. The conical egg chamber is at the top of the nest. Eggs leave the cloaca at intervals of 45 seconds and are caught by the hindlegs before being carefully placed in the egg chamber. The egg chamber is then closed and the nesting material pressed down firmly. The temperature in the egg chamber fluctuates between 26 and 33 °C. The incubation temperature influences sexual development; at temperatures below 29 to 30 °C more females are produced, and over 30 °C more males.

The female keeps the nest moist and protects it from predators. Whilst still in the egg, the hatchlings call their mother who uses her mouth to transport them to water after hatching. The hatchlings remain close to the mother for the first 3 to 4 weeks of life. Nevertheless, raccoons, otters, predatory fish, terrapins, and birds eat almost 80% of the young. A short time later, the family ties are broken and the young retreat to sheltered and secluded backwaters.

In the wild, young alligators feed on crayfish, snails, mussels, small fish, frogs, and large insects. Later they take larger prey such as fish, snakes, terrapins, birds, raccoons, and carrion. Many American Alligators are omnivorous and will eat food discarded by humans, such as potatoes, meat, sausage, and bread. Given the opportunity they will not hesitate to devour household pets. Of the 30 alligator farms in Florida, Ulrich (1989) reports that only five or six have bred these animals successfully and regularly. Other farmers buy eggs and hatchlings that are taken from the wild and reared in captivity. The first instance is called alligator "farming," the second alligator "ranching." These businesses incubate the eggs at around 33 °C, causing the young to hatch after around 64 days.

In the north of their distribution range Alligators hibernate for 2 to 3 months. In captivity this should also be allowed to stimulate breeding condition.

Alligator sinensis Fauvel, 1879
Chinese Alligator

Distribution: Eastern China: The lower reaches of the Yangtse-Kiang and its tributaries. Also in the provinces of Anhwei, Kiangsi, and Chekiang.
Description: The Chinese Alligator reaches a length of 1.40 to 2.10 m. Between the front corners of the eyes there is always a transverse ridge that is broken at the centre. The occipital scales at the rear of the head form two transverse rows, and on the neck there are pairs of nuchal scales lying behind one another. The dorsal scales form 6 longitudinal and 16 or 17 transverse rows. Adults are blackish brown whilst

Alligator sinensis

juveniles have whitish yellow transverse bands on both body and tail. These disappear with increasing age.

Habitat: Rivers, swamps with large reed beds, flooded areas with rich vegetation.

Lifestyle, husbandry, and reproduction: This is one of the most severely endangered species of Crocodilian. Their flesh and organs are supposed by the Chinese to possess all kinds of miraculous powers, and their skin is in great demand. Fishermen however regard the Chinese Alligator as a competitor and hunt them relentlessly. Habitat destruction is, however, the most important threat to the Chinese Alligator. Nowadays, state-owned farms are breeding these animals successfully.

Very little is known about the lifestyle of these creatures in the wild. They feed on various vertebrates, snails, mussels, frogs, fish, terrapins, birds, and mammals. They spend the winter in burrows which they dig themselves into mud on river banks in places that will not become flooded.

From other zoos, the Bronx Zoo in New York was able to obtain four Chinese Alligators which were housed in a fenced-off enclosure on the coastal swamps of southwest Louisiana. Here they hibernated from late October until early March in burrows which they dug themselves. In spring they were fed on salt-water fish and nutrias which were enriched with vitaminised minerals. Between 1977 and 1980 one female laid between 11 and 24 eggs each year in the second half of July. The eggs were laid in a nest built from plant material, and were jealously guarded and protected. A total of 29 young hatched between 1977 and 1980.

Caiman Spix, 1825
Caiman

The distribution range of the Caiman reaches from southern Mexico to the subtropics and tropics of South America.

The difference between the Caiman and the closely related genus *Paleosuchus* is that the Caiman has a raised, curved, transverse ridge between the front corners of the eyes ("spectacle rest"). There is no raised ridge along the back (as there is in *Melanosuchus*). Caiman species differ from *Paleosuchus* species by having at most three rows of tubercles at the back of the head and a yellowish green iris.

Caiman crocodilus (Linnaeus, 1758)
Spectacled Caiman, Crocodile Caiman

Distribution: Southern Mexico to subtropical South America. In Florida, colonies of introduced animals are surviving and breeding.
Description: From north to south, the short-snoutedness of this species increases. There are five teeth in each intermediate jaw. The branches of the lower jaw are firmly connected to one another up to the fourth and fifth tooth. The nuchal scales on the neck form two transverse rows, and the rear occipital scales on the back of the head four or five rows. Each of the front transverse rows has four scales whilst those at the rear of the neck have two. The dorsal scales form eight to ten longitudinal rows and 18 or 19 transverse rows. Males are larger than females with a broader head and tail as well as much stronger muscles in the neck. Like all Crocodilians, the Spectacled Caiman is capable of a physiological colour change. If they are relaxed they are light coloured, but when they are threatened they become dark (Wermuth, 1963, Trutnau, 1990). In addition to the nominate form, *Caiman c. crocodilus*, three subspecies are recognised: *Caiman c. apaporiensis, Caiman c. fuscus, Caiman c. yacare*.
Habitat: Slow-flowing rivers with muddy bottoms and rich vegetation. Backwaters of rivers, lakes, ponds, forest swamps, flooded fields with streams, and occasionally in brackish water swamps.
Lifestyle, husbandry, and reproduction: Since the end of the 1950s, the populations of Spectacled Caiman have been increasing in the wild in Florida. The first of these were vivarium specimens from Mexico and the north of South America. In the meantime a clutch of 37 *C. c. crocodilus* eggs was found, all of which hatched (Ulrich, 1989).

Caiman crocodilus

In the wild, the Spectacled Caiman will eat anything that it can overpower: water insects, crayfish, fish, amphibians, reptiles, birds, and small mammals as well as carrion. According to Trutnau (1990), a large number of toads was found in the stomach of a specimen of *Caiman crocodilus*. In captivity they will eat freshwater or saltwater fish, chicks, mice, rats, rabbits and the broken bones of slaughtered animals. In time, Caiman learn to distinguish their keeper from other humans.

In the breeding season, mating takes place both during the day and night. Shortly before the young hatch, the female opens the nest and carefully breaks the egg shells. The yellow to yellowish brown hatchlings have six or more broken transverse bands on the body and seven on the tail. The young remain close to their mother who protects them and emits a croaking sound when danger threatens.

Hirschfeld (1966) saw his Spectacled Caiman mating several times, the first time being at the end of January. The male clasps the female with the forelegs and pushes his tail beneath that of the female from the side. Around half of the body was on top of the female whilst the remainder was turned to one side. The female bent her tail sideways from the base to the tip. Copulation never lasted longer than 1 minute. The male Caiman frequently attacked a Nile Crocodile that was kept in the same enclosure and that had also attempted to mate with the female Caiman. The male Nile Crocodile had to be moved to a separate enclosure.

By early March, the abdomen of the female was considerably swollen and she attempted to dig on the land part of the enclosure. Using the hindlegs, sand and earth were pushed together to form a small mound, the size of which continued to grow as more nesting material was given.

At 1700 h on March 16, when the ground heating was in operation, the female, using her hindlegs, dug a hole some 25 cm deep and 35 cm in diameter in which to lay her eggs. The female raised her body on the hindlegs and pressed the flanks together. As the body was lowered, the first egg dropped into the nest. It took around 50 minutes to lay a clutch of 25 eggs. After a short pause, the eggs were carefully covered. Using the hindlegs, the female pushed together a mound around 25 cm high.

The female protected the eggs so well that the keeper was only able to remove 22 eggs by using a long-handled shovel. Of the 25 eggs laid, 23 were between 5.5 and 6.5 cm long. The remaining two eggs were double-yolked and 10 to 11 cm long. Both proved to be infertile. On May 18, the day before the heating was again switched on, a second Caiman female laid 23 eggs in a nest of earth, moss, and sand. The keeper bedded these eggs in moist, sterilised sand and covered them with moist, sterilised moss. Two bowls of water provided the necessary atmospheric humidity. A heating cable heated the ground to 26 to 28 °C. The air temperature above the bowls of water fluctuated between 35 and 40 °C and the humidity between 80 and 98%. At the start of the twelfth week of incubation, the female to whom the eggs belonged became very restless and aggressive towards the other occupants of the vivarium, especially when they approached the land part where the eggs had originally been laid. It was obvious that the female knew instinctively that hatching was imminent although there was no contact with the eggs.

At this time the female also had an enormous appetite. The first of the young hatched on August 8, the sixth and last on August 11. The remaining eggs were either addled or contained embryos which had died at an early stage in their development. One egg that was opened too early was found to contain a viable hatchling which died soon afterwards. The six young measured between 18 and 22 cm and weighed from 25 to 32.5 g. After only 8 months they had reached between 60 and 71 cm and weighed between 825 and 1350 g. Because of their differently arranged ventral scales it was always possible to identify each of the young individually from photographs. In the following year and in the same enclosure, 32 eggs were laid from which 24 young Caiman hatched (Hirschfeld, 1966, 1967).

Also in Stribrny's case (1978), the Caiman several times laid between 14 and 23 eggs. Success was only achieved, however, when the 6.1 × 3.9 cm eggs were placed in an incubator, the base of which was covered with sharp sand. This was covered in turn by a 10 cm layer of a sand/peat mixture in which the eggs were bedded. The eggs were covered with a further layer of peat and were incubated at temperatures from 28 to 32 °C and high atmospheric humidity. The young hatched after 98 to 104 days. They were between 15 and 18 cm long and weighed between 25 and 30 g. In

Caiman latirostris

winter it is not uncommon for Caimans to refuse food for 6 to 8 weeks, after which they will again feed greedily.

Caiman latirostris (Daudin, 1802)
Broad-Snouted Caiman

Distribution: Eastern Brazil, Uruguay, Argentina. There are probably individual populations in Paraguay and Bolivia.
Description: Representatives of this species may reach a length of 2 to 2.5, in exceptional cases up to 3 m. Because of the short and very broad snout, this species is easily distinguished from others. The small rear occipital scales on the back of the head form two transverse rows and the nuchal scales on the neck, four transverse rows. The foremost row consists of four scales and the next three, each of two scales. The dorsal scales form six to eight longitudinal and 18 or 19 transverse rows. Between the throat and the hindquarters, the unmarked ventral scales are arranged in 24 to 28 transverse rows. The scales on the flanks are lightly keeled. The back of the Broad-Snouted Caiman adults is a blackish, yellowish green or yellowish brown colour.

Habitat: These Caiman prefer calm, river backwaters and slow-flowing streams but may also be found in standing water or in brackish water swamps.
Lifestyle, husbandry, and reproduction: These Caiman breed at different times throughout their enormous distribution range. Females lay 20 to 80 eggs in a nest constructed from decaying plant material. The eggs measure around 7 × 4.5 to 7.0 cm. The nest is some 1.4 m in diameter and around 50 cm high. The eggs are incubated by the heat generated as this plant material decays even more. At a temperature of around 32 °C the young Caiman take around 86 days to develop and hatch.

Melanosuchus Gray, 1825

The sole species of this genus is the Black Caiman, *Melanosuchus niger*, from tropical South America. Black Caiman are the largest members of the Caiman group and in some exceptional cases may reach a length of up to 6 m. The more normal length, however, is 3 to 4.5 metres. In contrast to other Caiman species, the bony eye holes are elongated towards the nose. The surface of the upper lid is finely striped but not wrinkled. The flat, broad snout is quite long. Two rows of longitudinal scales form a wide, raised ridge along the centre of the back. The tubercles on the neck form at least four transverse rows.

Melanosuchus niger (Spix, 1825)
Black Caiman

Distribution. The Amazon region and central South America.
Description: Unlike other Caiman which have a greenish iris, the iris of the Black Caiman is light brown. The dorsal scales of this Caiman are arranged in 8 to 10 longitudinal rows and 18 or 19 transverse rows. Brock (1983) reported on the extraordinary colouring of a juvenile specimen some 50 cm long: The head and neck were grey with light olive and brown markings. The lower jaw was light with dark brown spots. The black colouring and transverse bands began immediately behind the neck tubercles. With increasing age these markings disappeared completely, producing a black adult with no markings on the light underside.
Habitat: Calm waters, slow-flowing waters, or standing lakes.
Lifestyle, husbandry, and reproduction: According to where in the large distribution range the animals live, the breeding season is from September to January. The female uses plant material to build a nest of 1.5 m in diameter and 80 cm high, in which 35 to 75 rough-shelled eggs are laid. The eggs measure around 9 × 5 to 6 cm. The female guards the eggs quite jealously. Frequently females build nests close to one another.

According to Brock (1983) the young are very intolerant of one another until they reach a length of 80 to 90 cm. As they grow larger they become more peaceful towards other smaller crocodiles until they reach sexual maturity. At that time they may control the entire water area for 7 or 8 weeks. Brock frequently heard the dull,

Melanosuchus niger

hoarse mating calls in the early hours of the morning. Black Caiman are sexually mature at 12 years of age.

Paleosuchus Gray, 1862
Smooth-Headed Caiman

Both species of this genus live in tropical South America. They are some of the smallest of the present-day crocodilia reaching a total length of only around 150 cm. In comparison to the genera *Caiman* and *Melanosuchus*, Smooth-Headed Caiman have only four teeth on each side of the intermediate jaw. The snout is relatively long and wedge-shaped. As opposed to other members of the family, the iris is not green, but chocolate brown. The common name "Smooth-Headed Caiman" refers to the fact that the transverse ridge between the rear corners of the eyes, so typical in other species, is absent in this species. The rear occipital scales on the back of the head form one or two transverse rows whilst the nuchal scales on the neck form up to a maximum of five transverse rows. The large, horny ventral scales are arranged in 17 to 19 transverse rows. A central row of ventral scales consists of 10 to 12 scales.

Paleosuchus palpebrosus (Cuvier, 1807)
Dwarf Caiman

Distribution: From Ituverava (Sao Paulo) westwards to Corumba (Mato Grosso) and Villa Maria on the upper Rio Paraguay. In the north through Columbia eastwards to British Guyana, Surinam, and Cayenne. Also at the mouth of the Amazon on the island of Mexicana (Lüthi & Stettler, 1977).

Description: The Dwarf Caiman is brown on the upper side. The underside is light with dark spots along the sides. Juveniles have dark spots or transverse bands along the back. The iris is chestnut brown and the upper eyelid is completely bony. The rear occipital scales on the back of the head form two transverse rows, and the nuchal scales on the neck four or five transverse rows that lie close together. The lower jaw has flecks and the sides of the snout fall away steeply creating a sharp angle. Their irregularly arranged dorsal scales form 6 to 8 longitudinal rows and 18 or 19 transverse rows. The maximum length is a little over 150 cm.

Habitat: A typical inhabitant of ecological niches. These Crocodilians live only in tropical rain forests in fast-flowing rivers, small streams, and in small river backwaters near fast-flowing water.

Lifestyle, husbandry, and reproduction: These nocturnal Caiman spend the daylight hours in deep, fast-flowing water. Lüthi (1983, 1986) reared a pair which were later housed in a 400 × 120 × 60 cm cellar enclosure. Approximately one-quarter of the enclosure was land. The enclosure was lighted for 12 hours daily using eight 40-W Tropical Daylight™ tubes. The water was heated to between 25 and 27 °C and was kept constantly clean by filtering (1500 1/h). The air temperature was kept at 26 to 28 °C.

The male was 1 m long and the female 10 cm shorter. Once weekly the animals were fed on newly killed mice, ox heart, or ox liver. Fish was always refused.

From mid-October the mating call of the male was heard regularly; it sounded like loud thunder and ended in a loud crackling noise. At this point the Caiman were given, once each week, a dead mouse that contained a Protovit™ capsule and a capsule of Ephynal™, a multivitamin and vitamin E preparation. The male gave his mating call throughout December, usually beginning in the early hours of the morning. As was the case each year, the Caiman refused food during December and January. The temperature was reduced by 10 °C and the lighting to 6 hours daily.

At the end of January, the female increased in girth but was still not eating. When the female started scratching on the land area, a large box of sand and several balls of hay were placed in the enclosure. On July 3, 1982, during the period from 1600 to 2300 h, nine eggs were laid, eight of which were bedded in a mixture of sand and wood shavings and were placed in an incubator with a temperature of 29 °C and 90% humidity. On the morning of October 15, one egg was found to be split and by midday the entire shell was broken off. Only the internal skin of the egg surrounded

become narrower towards the snout. The neck and dorsal scales are widely separated from one another. The latter form 16 transverse rows.
Habitat: Predominantly rivers.
Lifestyle, husbandry, and reproduction: Probably the same as other species. The Orinoco Crocodile is the species most threatened with extinction.

Crocodylus johnstoni Krefft, 1873
Australian Freshwater Crocodile

Distribution: Northern Australia from the Kimberley Region eastwards to northeast Queensland.
Description: This species grows to a length of 3.2 m and has an extremely narrow snout, the sides of which are almost parallel in the centre. The snout is at least 2.7 times as long as it is broad at the base. The six rows of neck scales and 19 transverse rows of dorsal scales are barely separated from one another. On each side of the intermediate jaw there are always five teeth. The snout is not wider at the height of the fifth tooth.

The grey or olive-brown upper side has dark irregular flecks, especially on the flanks. The underside is a whitish colour.
Habitat: This species is found on freshwater rivers and in lagoons, but may also be found on other waters.
Lifestyle, husbandry, and reproduction: Australian freshwater Crocodiles are diurnally active although they hunt for food mainly in the evening or at night. They feed on fish, frogs, crustaceans, smaller reptiles, birds, and mammals. At the end of the dry season (October to November), females lay up to 20 eggs in nests that they build on sandbanks.

Crocodylus mindorensis Schmidt, 1835
Philippine Crocodile

Distribution: The Philippine Islands of Luzon, Mindoro, Busuango, Masabate, Samar, Negros, Mindanao, Jolo, and Culion. Earlier occurrences on Masabate, Jolo, and Busuango have become extinct.
Description: The 1.80 to 2.40 m Philippine Crocodile was considered to be a subspecies of the New Guinea Crocodile, but in the Philippine Crocodile the wrinkles on the snout are more pronounced than in the other species. In 80% of these animals there are six rear occipital scales at the back of the head, whilst the remaining 20% have four or five rear occipital scales at the back of the head. On the light, unmarked underside there are 25 transverse rows of ventral scales. Adults are brown or dark olive with flecks. Juveniles are basically a yellowish brown with dark flecks or bands.
Habitat: Fresh and brackish water lakes, also on rivers and in swamps.
Lifestyle, husbandry, and reproduction: Very little is known about the lifestyle in the wild. It probably corresponds to that of *Crocodylus novaeguinea*.

Crocodylus acutus (above), Crocodylus johnstoni (below)

According to Gaulke (1986), this species used to be common on the flatlands of the Philippines; today however, only very few of the earlier populations still survive. The total population is thought to be between 500 and 1000 individuals. In a breeding station (Marine Laboratory, Damaguete City, Negros Oriental), Gaulke saw how the female of this species transports the young to the water in its mouth. The female checks the nest mound regularly and very carefully. When the first of the young begins to croak, it, or the egg, is carried individually to the water. On the west coast of Lake Naujan, newly hatched young are frequently found although adult animals are rarely seen. In some places females build their nest from fresh grass.

Crocodylus moreletii Duméril & Duméril, 1851
Morelet's Crocodile.

Distribution: Central America; East coasts of Mexico, Honduras, and Guatemala.
Description: These animals are considered to be fully grown at a length of around 2.5 m. The dorsal surface of these animals is very dark, indeed almost black. In front of the eyes there is an unpaired central, blunt, and oval swelling on the upper side of the snout. The dorsal scales are irregularly arranged. The short snout is rounded at the tip and only a little more than 1.5 times the basal breadth. The scales on the sides of the limbs are smooth.
Habitat: Diverse waters.
Lifestyle, husbandry, and reproduction: The same as all other species.

Crocodylus niloticus Laurenti, 1768
Nile Crocodile.

Distribution: Africa (with the exception of the extreme north) including Madagascar, Comoros, and Seychelles Islands.
Description: Usually 3 to 4 m long, in exceptional cases over 6 m. On the neck there are four to six rear occipital scales arranged in a transverse row. The dorsal scales are separated from the nuchal scales and are arranged in 6 to 8 longitudinal and 17 or 18 transverse rows. Seven subspecies are recognised. These are differentiated by the prominence of the neck collar; the various number of scales, and the various degrees of keeling and boniness of the scales.

On the upper side, Nile Crocodiles are dark olive. The unmarked underside is a porcelain colour. The light to dark grey, brown, or rarely maize-coloured juveniles have dark flecks and a slight hint of banding.
Habitat: In fresh and brackish water, in lakes, ponds, rivers, and streams; also in swamps mainly on the savannahs. Also found in the Central African rain forests.
Lifestyle, husbandry, and reproduction: In the wild, the Nile Crocodiles feed on fish, terrapins, and large mammals such as zebra, antelope, hyena, wild dogs, porcupine, young hippopotamus, lions, and even carrion.

Nile Crocodiles are sexually mature at 8 to 10 years old. During the breeding season males claim a territory which they defend against all other males. Mating usually takes place in the morning. During the breeding season, males and females stay close together. It is possible that at this time they live monogamously. The 16 to 80 eggs are laid around 5 months after mating and fertilisation. Using the hindlegs, the female digs a hole some 30 to 40 cm deep in which the eggs are laid and then covered with a mound of earth. The female guards the nest and does not feed for the 84 to 90 days that the eggs require to develop and hatch. As soon as the young begin to croak and signify that hatching is imminent, the female digs the young from the nest and carries them to the water in her mouth. The parents then protect the young from any possible predators. Nevertheless, many hatchlings fall prey to large fish, other crocodiles, snakes, herons, and especially Monitor Lizards. Other possible predators are hyena, mongoose, and other carnivorous mammals.

Because of numerous farms, the numbers of Nile Crocodiles have increased dramatically and in some countries they are no longer regarded as threatened.

In the spring of 1971, Jes (1981) received 13 specimens for the Cologne Aquarium from Rhodesia (now Zimbabwe). Within 10 years they had grown from 160–200 cm to 190–280 cm. In the Cologne Aquarium, the crocodiles are housed in an enclosure of 50 m^2 containing a large water section. A plant trough allows the animals to avoid sight of one another. In the winter the temperature is around 20 to 22 °C, in summer it increases to 25 to 30 °C. The food is mainly indigenous freshwater fish (roach, bream) with occasional rodents and poultry which is given twice weekly. In summer these animals eat around 10 kg each, whilst in winter they eat barely 500 g.

During the breeding season, from November to March, rivalry conflicts are quite common. The subordinate animals are forced to retreat to deeper water or another part of the enclosure. Mating usually takes place in January and February, occasionally also in November, December, and March, but always in the water. Because the land section was not suitable for egg laying, the eggs were inevitably laid in the water. Those that were discovered early enough were bedded in sphagnum moss and incubated at 28 to 30 °C and almost 100% atmospheric humidity. The young hatched after 100 to 108 days after having made croaking noises from inside the eggs. The young could be induced to croak by knocking on the sides of the incubator. On hatching the young had a length of 28 cm. Not quite 3 years later the juveniles had already reached a length of 140 cm. The young fed for the first time around 3 weeks after hatching. They ate small live fish from forceps, but quickly became accustomed to dead fish.

Crocodylus novaeguineae (Schmidt, 1828)
Freshwater Crocodile

Distribution: New Guinea.
Description: Freshwater Crocodiles may reach a length of 6 m but are usually much shorter. In front of each eye there is a bony longitudinal ridge, not so pronounced however as in the Saltwater Crocodile. The four rear occipital scales at the back of the head form a transverse row. Dorsal scales are arranged in 8 longitudinal and 16 or 17 transverse rows. Adult specimens are dark grey. Juveniles have dark flecks.

Habitat: This species prefers vegetation-rich and slow-flowing waters such as rivers, swamps, and swampy forests.
Lifestyle: In the wild, the Freshwater Crocodile feeds on various arthropods, amphibians, reptiles, and birds. When given the opportunity they will also eat small mammals. During the mating season, from July to October, the female builds a nest of plant material in the shade of a tree. The nest is around 50 cm high and 1.30 to 1.50 in diameter and usually contains around 30 eggs of 75 × 44 cm. The female guards the nest against predators and protects the young after hatching. At this time the young are around 25 to 30 cm and could easily fall prey to other animals. They are sexually mature at a length of around 2 m.

Crocodylus palustris Lesson, 1834
Mugger Crocodile

Distribution: The nominate form, the Mugger Crocodile (*C. palustris palustris*), lives in western Pakistan, India, Bangladesh, and Assam; The Ceylonese Mugger Crocodile (*C. palustris kimbula*) lives in Sri Lanka. The difference between the two subspecies is questionable.
Description: Only rarely do Mugger Crocodiles reach a length of 4 m. A more usual length is around 2.50 m. The most noticeable feature in comparison with the Nile Crocodile is the short, broad snout. The four to six large rear occipital scales at the back of the head are arranged in a transverse row, behind which a second transverse row of smaller tubercles is sometimes found. In the nominate form the dorsal scales form at most four longitudinal rows, whilst *Crocodylus palustris kimbula* has six. Moreover, in the nominate form, the two central rows of scales are broader than those at the sides. Adult Mugger Crocodiles are dark grey to dark olive in colour. The grey to light olive juveniles have dark flecks on both sides and dark rings on the tail.
Habitat: Rivers, jungle ponds, lakes, rice paddies, canals, and large fish ponds.
Lifestyle, husbandry, and reproduction: Whitaker & Whitaker (1983) breed the Mugger Crocodile on a farm in India. The main breeding group of one male and nine females live in an enclosure planted with Casuarines and trees to provide shade. The enclosure is equally divided into land and water, the depth of which during the monsoon season (November to January) is around 2.50 m, but which is normally around 1 m.

Females are sexually mature at 7 years of age whilst males require 9 years to achieve maturity. The females have become accustomed to making their nests in the sand and lay two clutches each year. The interval between each clutch is only around 1 month. This phenomenon has not been observed in any other species of crocodile. During the breeding season, from February to April, eight females built a total of 15 nests, each of which contained, on average, 26 eggs, from which a total of 236 young hatched, each around 26 cm long. At each feeding, the breeding group re-

Crocodylus novaeguineae (above), Crocodylus palustris (below)

quires 20 kg fish, 40 kg beef, and 40 kg bandicoots; thus each crocodile devours around 10 kg of food each month.

In the early days of rearing, the juveniles are fed mainly on crabs and finely ground meat. They are kept in groups of 30 specimens to each 2.50 × 2.50 m cement enclosure, 70% of which contains water around 30 cm deep.

Crocodylus porosus Schneider, 1807
Saltwater Crocodile

Distribution: India, Burma, Bangladesh, Sri Lanka, Kampuchea, Thailand, Vietnam, Philippines, Palau Islands, Malaysia, Singapore (thought now to be extinct), Indonesia, Brunei, Papua New Guinea, Solomon Islands, Australia, and Vanuatu.

Description: Saltwater Crocodiles may reach a length of almost 10 m, although in the wild, creatures of this length are extremely rare (Jungnickel, Sommerlad & Trutnau, 1991). The relatively long snout is more than twice as long as the width at the base. In front of each eye there is a tubercular, bony longitudinal ridge that runs down to the nasal tubercles. Juveniles have five teeth in the intermediate jaw, whilst adults have only four. The branches of the lower jaw are joined together up to the fourth or fifth tooth. The rear occipital scales on the back of the head are either absent or only barely visible, especially in specimens from Sri Lanka. The elliptical dorsal scales are significantly separated from the neck scales. They form 6 to 8 longitudinal rows and 16 or 17 transverse rows. Only the centre of each scale is bony. Adult Saltwater Crocodiles are uniform light olive in colour.

Habitat: Brackish water swamps, at river mouths, and in the open sea. More rarely in inland rivers. Specimens in Sri Lanka are less frequently found in the sea and live predominantly in fresh water.

Lifestyle, husbandry, and reproduction: Juveniles eat insects, crayfish, fish, and smaller reptiles. Adults eat larger animals, water birds, large fish and terrapins, wild pigs, deer, apes, and even household pets such as cats and dogs, goats, cattle, and horses. The Saltwater Crocodile does not only catch its prey whilst drinking on the banks, but also hunts for prey on land during the hours of darkness.

There are countless farms that breed Saltwater Crocodiles. Males reach sexual maturity at the age of 12 years when they are around 3 m long. Females are sexually mature at 8 years of age, with a body length of 2 m. Since these animals can live to 80 years of age, females are theoretically able to lay over 2,000 eggs during their lifetime.

In eastern India, mating takes place in February or March, after which females use plant material to build a nest 50 to 90 cm high and 1.2 to 2.5 m in diameter in which the eggs are laid. In northern Australia, Saltwater Crocodiles breed from November to March. Obviously the breeding season is governed by the end of the regional dry period.

The 20 to 90 eggs, each measuring around 7.5 × 5.0 cm are placed in the nest around 25 cm below the surface where temperatures of 27 to 33 °C prevail. The 25 to 30 cm young hatch after 78 to 90 days. In the wild they remain close to the mother for some 10 weeks. Most however, fall victim to predators. Of the 1,000 juveniles at the Samutprakarn Farm in Thailand, three to five are light skinned.

In almost all crocodile species variations in colour invariably occur. An extreme example is "the first white crocodile in Europe" (Brock, verbal communication), a *C. porosus porosus* which has been kept at Wilhelma Zoo in Stuttgart, Germany

Crocodylus porosus

since 1967. In 25 years (up to 1993), it had reached a length of almost 5 m. It is not albinotic, but rather a whitish yellow with small and large, deep black flecks over the entire body from the forelegs to the tip of the tail.

In 1975, Brock (written communication) was offered a 35 cm Saltwater Crocodile. However, because of its eventual size, this offer was refused. These animals can only be adequately housed in a large greenhouse complex with a large water section, which is only possible in zoos or other similar establishments because these animals are always ready to attack and bite without warning.

Crocodylus rhombifer Cuvier, 1807
Cuban Crocodile

Distributiom: Central Cuba and the Isle of Pines.
Description: In front of the eyes there is a raised, triangular area. The dorsal scales are fairly regularly arranged. Each intermediate jaw has five teeth. The black back

Crocodylus rhombifer

has yellow flecks. Cuban Crocodiles are usually 2.50 to 2.80 m long. They may, however, grow a little longer.
Habitat: Fresh water.
Lifestyle, husbandry, and reproduction: As in all other species.

Crocodylus siamensis Schneider, 1801
Siamese Crocodile

Distribution: Far India (Thailand and Indochina), Sunda Archipelago (Java and Borneo).
Description: In front of the eyes there is a raised, triangular area. The dorsal scales are fairly regularly arranged. The snout of this species is conspicuously flat. Adults have only four teeth in the intermediate jaw (in front of the side indentation at the tip of the snout). Between the eyes there is a raised, longitudinal ridge. Siamese Crocodiles reach a length of 3.30 to 3.80 m.
Habitat: Fresh water.
Lifestyle, husbandry, and reproduction: As in all other *Crocodylus* species. The Siamese Crocodile has been bred for many years on farms.

Osteolaemus Cope, 1861

The only species of this genus, the Dwarf Crocodile, lives in west and southwest Africa, southwards to Angola. Externally the Dwarf Crocodile is reminiscent of an alligator in having a short snout and no scale combs on the limbs. At the centre of the nasal cavity there is a bony separating plate. The external nasal openings are separated by a wide groove.

Osteolaemus tetraspis Cope, 1861
African Dwarf Crocodile

Distribution: West and Central Africa (Sierra Leone, Guinea, Ghana, Togo, Nigeria, Cameroon, Gabon, Angola, Liberia, northeastern Congo).
Description: Usually 1.30 to 1.50 m long, rarely up to 2 m. The short snout has neither ridges nor swellings. Adults have four teeth in the intermediate jaw, whilst juveniles have five. The rear occipital scales at the back of the head are arranged in one or two transverse rows. The pairs of nuchal scales are close together. The dorsal scales form 6 to 8 longitudinal rows and 17 to 20 transverse rows. Adults are dark brown to black. Juveniles have many transverse bands on a dark background. On the underside these animals are black, or have at least some black flecking. Dwarf Crocodiles have a distinct collar and heavily keeled, bony scales on the flanks.
Habitat: Predominantly in dense rain forests and in small rivers, streams, and ponds.
Lifestyle, husbandry, and reproduction: Dwarf Crocodiles frequently wander far away from water. They usually lead an individual and very secretive lifestyle. Because they are active at dusk and during the night, very little is known about their lifestyle in the wild. The daylight hours are usually spent in a burrow near the banks of the water. Fish and amphibians form the main part of their diet. Dwarf Crocodiles often fall prey to Nile Crocodiles which occur in the same areas. It is perhaps for this reason that they are mainly active after dark.

Helfenberger (1981) kept one male of 140 cm and one female of 130 cm, together with two alligators in an enclosure with a ground areas of 2.5 m^2. The enclosure was lined with sheets of slate and contained a 60 × 60 cm water section with water 30 cm deep. The water temperature fluctuated between 23 and 30 °C.

At the end of October 1977, repeated mating attempts were seen over a period of 11 days, always in the evening. Mating calls were also heard at irregular intervals. The male clasped the female in his forelegs and both animals turned slightly to one side. At the end of March the female refused all food and became more and more aggressive. Despite the hard ground the female began scratching on May 14, 1978, and laid eggs on May 24. While laying eggs, the female stood with the forelegs on the back of an alligator. Within 20 minutes, 14 eggs had been laid on the stone ground. The eggs measured 6.5 × 3.4 cm and weighed 46.8 g. They were bedded 3 to 5 cm deep in a mixture of sand and earth and were placed in an incubator. At temperatures between 27 and 33 °C and atmospheric humidity of 85 to 90%, the first two young hatched after 85 days. Because croaking calls had been heard the keeper had opened the eggs during the previous day. From the second day after hatching, the young ate macropods, crickets, and earthworms. They "indicated" that they were hungry by "croaking."

Richter (1981) received 7 from 16 eggs that a female had laid in a sand pit. These were bedded in a bucket containing a mixture of peat and sand. The bucket was placed in an aquarium containing 2 cm of water. The aquarium was almost completely covered by a sheet of glass, leaving only a very small gap for ventilation. During the day the temperature in the room where the eggs war placed was 30 °C, falling at night to 26 °C. The atmospheric humidity fluctuated between 70 and 90%.

At the same time, the owner of the animals attempted to incubate seven eggs from the same clutch in a peat-filled aquarium placed in the greenhouse containing his crocodiles, i.e., under the same climatic conditions as the parents, a 150 cm male and a 130 cm female. The 170 × 230 × 100 cm water section contained water only 25 cm deep. The land section measured 230 × 100 cm. In summer, the temperatures varied between 20 and 32 °C. After 4 months, the eggs showed no indication of embryonal development. The eggs that Richter was incubating, however, contained three live young after 130 days. After the eggs had been opened the young emerged and began to croak loudly. The smallest of the young (22 cm, 44g) died 3 days later. The remaining two young weighed 52 g and were 22.5 cm long. Because of the residual umbilical cord, the young were only placed in an aqua vivarium with land and water sections five days after hatching. They were given ultraviolet light from an Osram-Vitalux™ lamp for 10 minutes daily. After several days they fed on small grasshoppers and guppies (*Poecilia reticulata*). Later they took pink mice and small pieces of fish from long forceps.

Tomistoma Müller, 1846

The sole representative of this genus is the False Gharial (*Tomistoma schlegelii*), the home of which is in the extreme southwest of eastern India. Because of its long narrow snout, this species is reminiscent of a member of the genus *Gavialis*.

Tomistoma schlegelii Müller, 1838
False Gharial, Sunda Gharial

Distribution: Southern Thailand, Malaysia, and the two Sunda Islands, Sumatra and Borneo.

Description: Only rarely does this species reach a length of over 5.50 m. A more usual length is a little over 3 m. In contrast to the Ganges Gharial, the snout of the False Gharial is not particularly well set-off at its base from the main part of the skull. The small rear occipital scales at the back of the head form two transverse rows. There is almost no differentiation between the nuchal scales and the dorsal scales which form 6 longitudinal rows and 22 or 23 transverse rows.

On a nut-brown background there are black flecks which frequently form transverse bands on the tail. Juveniles also have transverse bands on the body.
Habitat: Rivers and swamps.
Lifestyle: The lifestyle of this species has not been investigated to any great extent. Its diet consists essentially of fish, so that in captivity it must be fed mainly on freshwater fish. Brock (written communication) also feeds his *Tomistoma* on dead mice and strips of beef heart. In captivity these crocodilians easily tend to rachitic deformations that usually cause the snout to curve to an arc shape. This condition is usually caused by inadequate feeding and housing. The False Gharial has not yet been bred in captivity. Almost nothing is known about their breeding habits in the wild. It is thought that they make simple nests using some plant material. The False Gharial is less tied to water than the Ganges Gharial.

GHARIALS
Family Gavialidae

Gavialis Oppel, 1811
Gharial

In earlier times there were countless Gharials living on the river systems of the Indus, Ganges, Brahmaputra, Kolawadi, and Iriwadi. Today however, they are almost extinct because they are relentlessly hunted for their skin and because they are regarded as competitors for food. River regulation has also caused enormous habitat destruction.

Nowadays, the Ganges Gharial is bred in large numbers on many farms and each year a large quantity of juveniles are released into the wild once they have reached a length of around 1.50 m. They are released into nature reserves where they are safe from hunters.

The skull of these animals is greatly elongated. The snout is very narrow and significantly set-off from the main part of the head. The jaws are almost identical in shape. All teeth fit in-between one another and point slightly outwards.

Gavialis gangeticus (Gmelin, 1789)
Ganges Gharial

Distribution: Western Pakistan, India, Bangladesh, Burma.
Description: Normally these animals "only" reach a length of 4.50 m, but in extremely rare cases they may attain a length of 7 m. At the back of the head there is only one pair of rear occipital scales. On the upper side, the dorsal scales form 6 longitudinal rows and 21 or 22 transverse rows. Adults are dark olive whilst the

Osteolaemus tetraspis (above), Gavialis gangeticus (below)

young are lighter with dark transverse bands. Males may be immediately recognised by the large bulbous swelling on the end of the snout.

Habitat: See above.

Lifestyle: Of all crocodilians living at present, the Ganges Gharial is the species that is most tied to water. On land, the relatively poorly developed limbs propel the stout body very clumsily. Gharials come onto land only to bask and lay eggs.

Female Gharials become sexually mature at 11 years old, when they attach themselves to a "harem" of four to six other females and one male. Gharials mate during November and December. In March, every fertilised female begins to dig several nests on sandbanks where they will be safe from flooding. The eggs, which are around 9 cm long and take 60 to 80 days to hatch, are guarded and protected by the female throughout their incubation.

Bibliography

Abraham, G. (1980): Das Portrait *Varanus acanthurus.*—Sauria, 1 (2): 2.
Arnold, E. N. (1989): Towards a phylogeny and biogeography of the Lacertidae: Relationsships within an Old-World family of lizards derived from morphology.—Bull. brit. Mus. nat. Hist., London, 25 (2) 209-257.
Baumann, F. (1913): Reptilien und Batrachier des Berner Naturhistorischen Museums aus dem Battak-Gebirge von West-sumatra.—Zool. Jb. Syst., Jena, 34: 257-278.
Bech, R. & U. Kaden (1991): Vermehrung von Terrarientieren. Echsen.—Urania-Verlag, Leipzig, Jena, Berlin.
Becker, C. (1982): Einige Bemerkungen zur Fortpflanzungsbiologie der Smaragdeidechse (*Lacerta viridis*) (Reptilia: Sauria: Lacertidae).—Salamandra, 18 (3/4): 121 bis 137.
Behrmann, H. J. (1981): Haltung und Nachzucht von *Varanus t. timorensis* (Reptilia: Sauria: Varanidae).—Salamandra, 17 (3/4): 198-201.
Beutler, A. & U. Heckes (1986): *Podarcis milensis* (Bedriaga, 1882)—Milos-Eidechse.—In: Böhme, W. (Hrsg.): Hb.d.Reptilien und Amphibien Europas: 142-154.
Bischoff, W. (1970): *Lacerta strigata* Eichwald, 1831, eine schöne und interessante Eidechse.— Aqu. Terr., Leipzig, Jena, Berlin, 17, (2): 48-49.
—— (1976): Echsen des Kaukasus 10. Die Streifeneidechse, *Lacerta strigata*, Eichwald, 1831.—Aqu. Terr. Leipzig, Jena, Berlin, 23: 84-88.
—— (1978): Beiträge zur Kenntnis der Echsen des Kaukasus.—Salamandra, 14, 4: 178-202.
—— (1981): Freiland- und Terrarienbeobachtungen an der Omaneidechse, *Lacerta jayakari*, Boulenger, 1887 (Reptilia: Sauria: Lacertidae).—Zeitschr. Kölner Zoo, 24 (4): 135-143.
—— (1981): Bermerkungen zur Herpetofauna der Vereinigten Arabischen Emirate, insbesondere zur Omaneidechse (*Lacerta jayakari*).—herpetofauna, 11 (März): 12-16.
—— (1984): *Lacerta agilis* Linnaeus 1758—Zauneidechse.—In: Böhme, W. (Hrsg.):Handbuch der Reptilien und Amphibien Europas: 23-68.
—— (1985a): Die Herpetofauna der Kanarischen Inseln III. Die Skinke der Gattung *Chalcides.*—herpetofauna, 7 (36): 13-21.
—— (1985b): Die Herpetofauna der Kanarischen Inseln IV. Die Atlantische Eidechse, *Gallotia atlantica* (Peters & Doria, 1882).—herpetofauna, 7 (37): 15-24.
—— (1985c): Die Herpetofauna der Kanarischen Inseln V. Die Rieseneidechsen der Gattung *Gallotia.*—herpetofaun, 7 (38): 11-21.
—— (1985d): Die Herpetofauna der Kanarischen Inseln VI. Die Kanareneidechse, *Gallotia galloti* (Oudart, 1839).—herpetofauna, 7 (39): 11-24.

— (1985e): *Lacerta* (Timon) *pater* Lataste. — Amph./Rept.-Kartei, 29-30—Beilage in Sauria, Berlin-W., 7 (4).
— (1985f): *Lacerta* (Timon) *lepida* Daudin.—Amph./Rept.-Kartei, 19-24—Beilage in Sauria, Berlin-W., 7 (3).
— (1986): *Podarcis filfolensis* (Bedriaga, 1876)—Malta-Eidechse.—In: Böhme, W. (Hrsg.): Handbuch der Reptilien und Amphibien Europas: 50-64.
— (1987): *Lacerta* (Omanosaura) *jayakari* Boulenger—Amph./Rep.-Kartei, Beilage in Sauria, Berlin-W., 9 (11): 65-70.
— (1991): Übersicht der Arten und Unterarten der Familie Lacertidae 1. Die Gattungen *Acanthodactylus, Adolfus, Algyroides* und *Australolacerta*.—Die Eidechse, 1: 18-23.
— (1991): Übersicht der Arten und Unterarten der Familie Lacertidae 2. Die Gattungen *Eremias, Gallotia, Gastropholis, Heliobolus, Holaspis*, und *Ichnotropis*.—Die Eidechse, 2: 14-21.
— (1991): Übersicht der Arten und Unterarten der Familie Lacertidae 3. Die Gattung *Lacerta*.—Die Eidechse, 3: 5-16.
— (1991): Übersicht der Arten und Unterarten der Familie Lacertidae 4. Die Gattungen *Latastia, Meroles, Mesalina, Nucras, Ophisops, Pedioplanis* und *Philochortus*.—Die Eidechse, 4: 17-25.
— (1992): Übersicht der Arten und Unterarten der Familie Lacertidae 5. Die Gattung *Podarcis*.—Die Eidechse, 5: 6-20.
— (1991): Übersicht der Arten und Unterarten der Familie Lacertidae 6. Die Gattungen *Poromera, Psammodromus, Pseuderemias, Takydromus* und *Tropidosaura*.—Die Eidechse, 6: 13-17.
Bohn, H. J. (1987): Zur Haltung und Vermehrung des Scheltopusiks (*Ophisaurus apodus thracius*).—Elaphe, 9 (3): 41-42.
Böhme, W. (1981) (Hrsg.): Handbuch der Reptilien und Amphibien Europas. Band 1, Echsen (Sauria) I. (Gekkonidae, Agamidae, Chamaelonidae, Anguinidae, Amphisbaenidae, Scincidae, Lacertidae I).—Akademische Verlagsgesellschaft. Wiesbaden.
— (1984) (Hrsg.): Handbuch der Reptilien und Amphibien Europas, Band 2/I, Echsen (Sauria) II, (Lacertidae II, *Lacerta*).—AULA-Verlag, Wiesbaden.
— (1986) (Hrsg.): Handbuch der Reptilien und Amphibien Europas. Band 2/II, Echsen (Sauria) II, (Lacertidae III: *Podarcis*).—AULA-Verlag, Wiesbaden.
— (1986): *Podarcis waglerianna* (Gistel, 1868)—Sizilianische Mauereidechse.—In: Böhme, W. (Hrsg.): Handbuch der Reptilien und Amphibien Europas: 377-387.
— (1988): Zur Genitalmorphologie der Sauria: funktionelle und stammesgeschictliche Aspekte.—B. zool. Monogr. 27: 1-76.
— (1988): Der Arguswaran (*Varanus panoptes*, Storr, 1980) auf Neuguinea: *V. panoptes horni* ssp. n. (Sauria: Varanidae).—Salamandra, 24 (2/3): 87-101.
— (1989): Zur systematischen Stellung der Amphisbänen (Reptilia: Squamata), mit besonderer Berücksichtigung der Morphologie des Hemipenis.—Z. zool. Syst. Evol. forsch., 27: 330-337.
— (1991): The Identy of *Varanus gouldii* (Gray, 1838), and the Nomenclature of the V. *gouldii*-Species Complex.—Mertensiella, 2: 38-41.
— , J. P. Fritz & F. Schütte (1987): Neuentdeckung einer Großechse (Sauria: Varanus) aus der Arabischen Republik Jemen.—herpetofauna, 9 (46): 13-20.

Bosch, in den, H. A. J. (1983): Voorplantings gegevens van *Algyroides moreoticus*, Bibron & Bory, 1833, de Peleponnesos kielhagedis.—Lacerta, Den Haag, 41 (10/11): 182-194.

—— (1986): Zu Fortpflanzung und sozialem Verhalten von *Psammodromus hispanicus*, Fitzinger, 1826, nebst einigen Bemerkungen zu *Psammodromus algirus* (Linnaeus, 1766) (Sauria: Lacertidae).—Salamandra, 22 (2/3): 113-125.

—— (1988): Erste Daten zum Fortpflanzungsmodus von *Ophiomorus punctatissimus* (Bibron & Bory, 1833) (Sauria: Scincidae).—Salamandra, 24 (1): 53-58.

Branch, B. (19884): Field guide to the snakes and other reptiles of Southern Africa.—Cape Town (Struik), 326 S.

Breustedt, A. (1990): Erfahrungen bei der Haltung und Nachzucht von *Zonosaurus karsteni* (Grandidier, 1869).—Sauria, Berlin, 12 (3): 3-7.

Bringsoe, H. (1986): *Podarcis peloponnesiaca* (Bibron & Bory, 1833)—Peloponnes-Eidechse. — In: Böhme, W. (Hrsg.): Handbuch der Reptilien und Amphibien Europas: 209-230.

Busack, S. D. (1978): Diurnal surface activity in the amphisbaenian *Blanus cinereus* (Vandelli), 1797) (Reptilia, Lacertilia, Amphisbaenidae). J. Herpetol. 12 (3): 428.

Cei, J. M. (1986): Reptiles del Centro Centro-oste y sur de la Argentina.—Museo Regionale di Science Naturali Torino.

Chaumont, F. (1963): Meine Beobachtungen bei der Geburt kleiner Blauzungenskinke.— DATZ, 16: 151-152.

Cooper, J. S. (1958): Observations on the eggs and young of the Wall Lizard (*Lacerta muralis*) in captivity.—Brit. J. Herpetol., London, 2: 112-121.

—— (1965): Notes on fertilisation, the incubation period and hybridisation in *Lacerta*.—Brit. J. Herpetol., London, 3: 218-220.

Cyren, O. (1934): Die alte Brückenechse in Upsala.—Bl. Aqu. u. Terr., 214-216.

Daele, F. van (1980); Geslaade kweek van *Mabuya breyicollis* in het terrarium.—Lacerta, Den Haag, 38 (9): 85-87.

Darewskij, I. S. (1984): *Lacerta strigata* Eichwald, 1831—Kaspische Smaragedeidechse.—In: Böhme, W. (Hrsg.): Handbuch der Reptilien und Amphibien Europas: 82-99.

Deraniyagala, P. E. P. (1953): A colored atlas of some vertebrates from Ceylon. Vol. II: Tetrapod reptilia. Colombo (Ceylon Gov. Press), i-xii, pls. I–XI, 1-35 44 figs., 100 pp.

Doria, G. (1888): Note Erpetologische—Alcuni nouvi Sauri raccolti in Sumatra dal Dr. O. Beccari.—Ann. Mus. civ. stor. nat. Genova (2), 6: 646-642, 1 pl.

Duinen, J. J. van (1983): Varanenkweek in het Noorderdierepark to Emme.—Lacerta, Den Haag, 42 (1): 12-14.

Eichenberger, P. (1981): Smaragdeidechse kweke op bestelling.—Lacerta, Den Haag, 39 (6/7): 72-76.

Eidenmüller, B. (1986): Beobachtungen bei der Pflege und Nachzucht von *Varanus* (Odatria) *t. timorensis* (Gray, 1831) (Sauria: Varanidae).—Salamandra, 22 (2/3): 157-161.

—— (1990): Beobachtungen bei der Haltung und Nachzucht von *Varanus* (Varanus) *mertensi*, Glauert, 1951.—Salamandra, 26 (2/3): 132-139.

—— & H. G. Horn (1985): Eigene Nachzuchten und der gegenwärtige Stand der Nachzuchten von *Varanus* (Odatria) *storri*, Mertens, 1966 (Sauria: Varanidae).—Salamandra, 21 (1): 55-61.

—— & R. Wicker (1991): Einige Beobachtungen bei der Pflege und Nachzucht von *Varanus* (Odatria) *timorensis similis*, Mertens, 1958.—Salamandra, 27 (3): 187-193.

—— & R. Wicher (1992): Über eine Nachzucht von *Heloderma suspectum* (Cope, 1869).—Salamandra, 28 (2): 106-111.
Eikhorst, R. & W. Eikhorst (1982): Zur Fortpflanzung der Spanischen Kieleidechse (*Algyroides marchi*, Valverde, 1958) (Reptilia: Sauria: Lacertidae).—Salamandra, 18 (1/2): 56-64.
Eikhorst, W., R. Eikhorst, H. K. Nettmann, & S. Rykena (1979): Beobachtungen an der Spanischen Kieleidechse *Algyroides marchi*, Valverde, 1958 (Reptilia: Sauria: Lacertidae).—Salamandra, 15 (4): 254 bis 263.
Erdfelder, K.-H. (1984): Haltung und Zucht des Stachelschwanzwarans *Varanus acanthurus*, Boulenger, 1885. — Sauria (1): 9-11.
Esterbauer, H. (1958): Bermerkungen zum Lebensraum und Verhalten der Glattechse *Mabuya vittata* in Syrien.—DATZ, 38: 416-419.
—— (1986): Zur Lebensweise von *Eumeces schneideri pavimentatus* in Syrien.—DATZ, 39: 227-278.
Ferguson, M. W. J. & T. Joanen (1982): Temperature of egg incubation determines sex in *Alligator mississippiensis*.—Nature, London, 216 (5860): 850-853.
—— (1983): Temperature —dependent sex determination in *Alligator mississippiensis*.—J. Zool., London, 200: 143-177.
Flugi, U. (1990): Bericht über die Haltung und Nachzucht des Storrschen Zwergwarans (*Varanus storri*, Mertens, 1966).—herpetofauna, 12 (67): 31-34.
Forman, F. & B. Forman (1981): Herpetologische Beobachtungen auf Korsika.—herpetofauna, 3 (10:) 12-16.
Franz, W., U. Bachmann & R. Braun B999 (1976): Außergewöhnliche Todesfälle durch Amöbiasis bei einer Brückenechse (*Sphenodon punctatus*), bein jungen Suppenschildkröten (*Chelonia mydas*) und bei einer Unechten Karettschildkröte (*Caretta caretta*). I. Amöbiasis bei *Sphenodon punctatus*.—Salamandra, 12 (2): 94-102.
Franzen, M. (1986): Herpetologische Beobachtungen im Santa-Rosa-Nationalpark, Costa Rica.—herpetofauna, 8 (41): 24-33.
—— (1986): Zur winterlichen Aktivität einiger Echsen in der südlichen Türkei.—herpetofauna, 8 (45): 6-10.
—— (1990): Die Eidechsenfauna (Lacertidae) der Türkei.—Die Eidechse, 1:3-9.
—— (1991): Beobachtungen zur phytophagen Ernährung von *Lacerta rudis* und *Lacerta clarkorum*.—Die Eidechse, 2: 22-23.
—— & U. Heckes (1992): Zum Vorkommen der Wieseneidechse *Lacerta praticola*, Eversmann, 1832, in der europäischen Türkei.—Salamandra, 28 (2): 129-137.
Freitag, W. (1983): Meine Erfahrungen bei der Haltung der Riesensmaragdeidechse *Lacerta trilineata*.—Elaphe, 5 (1): 10-11.
—— (1983): Meine Erfahrungen mit *Platsaurus guttatus*.—Elaphe, 5 (3): 37-38.
Froesche, P. (1980): Zucht und Aufzucht von *Lacerta viridis*, der Smaragdeidechse.—Das Aquarium, 134: 428-432.
—— (1983): Aggressive Kleinechse *Podarcis melisellensis*.—Das Aquarium, 166: 212-216.
Frommer, J. (1984): Nachzucht von *Podarcis pityusiensis*.—Elaphe, 6 (1): 8-9.
Fuchs, K. H., Mertens, R. & Wermuth, H. (1974): Zum Status von *crocodylus cataphractus* und *Osteolaemus tetraspis*.—Stuttgart. Beitr. Naturk., (A) 266: 1-8.
—— (1974): Die Unterarten des Nilkrokodils, *Crocodylus niloticus*.—Salamandra, 10 (3/4): 107-114.

Gaulke, M. (1986): Beitrag zur Kenntnis des Philippinen-Krokodiles, *Crocodylus mindorensis*.—herpetofauna, 8 (40): 21-26.

—— (1986): Über die Situation des Bindeswarans (*Varanus salvator nuchalis*) auf Negros, Philppinen.—herpetofauna, 8 (44): 16-18.

Gericke, F. (1983): Corytophanes cristatus (Merrem, 1820) ein seltener Gast aus Mittelamerika.—Sauria, 5 (1): 21-24.

Grantsau, R. (1966): Enyalius catenatus, das brasilianische „Cameleon".—DATZ, 19: 217-219.

Gross, J. (1989): Pflege, Geschlechtsbestimmung und Zucht der Tannenzapfenechse.—DATZ, 42: 612-613.

Grossmann, W. (1986): Erste Erfahrungen bei der Haltung und Nachzucht des Olivfarbenen Baumskinkes *Dasia olivacea*, Gray, 1838.—Sauria, 8 (4): 13-21.

Gruber, U. (1986): *Podarcis erhardii* (Bederiaga, 1876)—Ägäische Mauereidechse.—In: Böhme, W. (hrsg.): Handbuch der Reptilien und Amphibien Europas: 25-49.

—— (1986): *Podarcis gaigeae* (Werner, 1930)—Skyros-Mauereidechse.—In: Böhme, W. (Hrsg.): Handbuch der Reptilien und Amphibien Europas: 65-70.

Grundke, F. D. & B. Grundke (1992): Feldherpetologische Beiträge zur Fauna Australiens. Teil 2: Beobachtungen an *Varanus varius* in New South Wales und Queensland, unter besonderer Berücksichtigung der gebänderten Farbphase.—herpetofauna, 14 (80): 11-22.

Gruschwitz, M. & W. Böhme (1986): *Podarcis muralis* (Laurenti, 1768)—Mauereidechse.—In: Böhme, W. (Hrsg.): Handbuch der Reptilien und Amphibien Europas: 155-208.

Häberle, H. (1973): Ersatzfutter für den Panzerteju (*Dracaena guianensis*).—Das Aquarium, 49: 291.

—— (1979): Der Panzerteju, *Dracaena guidensis*.—DATZ, 32: 28-29.

Hämmerli, B. (1986): Geburt eines Keilkopf-Kaimans (*Paleosuchus trigonatus*) Daten, Fakten, Erfahrungen und Empfehlungen.—DATZ, 40: 129-132.

—— (1987): Geburt eines Keilkopf-Kaimans (*Paleosuchus trigonatus*) Daten, Fakten, Erfahrungen und Empfehlungen.—DATZ, 40: 176-178.

Harbig, P. (1986): Haltung und Zucht des gefleckten Walzenskinks *Chalcides ocellatus* (Forskal, 1775).—Sauria, 8 (3): 7-9.

—— (1988): Erste Beobachtungen bei der Haltung und Nachzucht des Khoratskinks *Riopa koratense* (Smith, 1917).—*Sauria, 10 (3):* 3-6.

Hart, H. (1978): Australische Immigranten: Stachelschwanzskinke. Pflege und Zucht von *Egernia cunninghami*.—Aquarien-Mag., 12: 534-536.

Hartmann, U. K. (1989): Beitrag zur Biologie des Apothekerskinks *Scincus scincus* (Linnaeus, 1758).—herpetof., 11 (59): 17-25.

—— (1989): Beitrag zur Biologie des Apothersknks *Scincus scincus* (Linnaeus, 1758). Teil 2: Apothekerskinke im Terrarium.—herpetofauna, 11 (60): 12-24.

Hauschild, A. (1986): Bemerkungen zur Haltung und Zucht der Langschwanzeidechse *Takydromus sexlineatus ocellatus*, Cuvier, 1829.—herpetofauna, 8 (44): 11-14.

—— (1988): Bemerkungen zu Haltung und Zucht des Schneckenskinkes *Tiliqua gerrardii* (Gray, 1845) (Sauria: Scincidae).—Salamandra, 24 (4): 248-257.

Heck, L. (1955): Der Kap-Waran (*Varanus albigularis* Daud.).—DATZ, 8: 190 bis 191.

Helfenberger, N. (1981): Ein Beitrag zur Fortpflanzungsbiolgie von *Osteolaemus t. tetraspis*.—herpetofauna, 11: 9-11.

Henderson, R. W. & A. Schwartz (1984): A guide to the identification of the amphibians and reptiles of Hispaniola.—Special Publ. Biol. Geol. Nu. 4, Milwaukee, Public Mus.: 1–70.

Hendrickson, J. (1966): Observations on the fauna of Pulau Tioman and Palua tulai. 5. The reptiles.—Bull. Raffles Mus., Singapore, 34: 53–71.

Henle, K. (1981): Die Herpetofauna Neuseelands, Teil 1: Auf der Suche nach Brückenechsen und altertümlichen Fröschen.—herpetofauna, 12: 9–12.

—— (1983): Auf der Suche nach Brückenechsen und Urföschen.—DATZ, 36: 74–76.

—— & Ch. J. J. Klaver (1986): *Podarcis sicula* (Rafinesque-Schmalz, 1810)—Ruineneidechse.—In: Böhme, W. (Hrsg.): Handbuch der Reptilien und Amphibien Europas: 254–342.

Hilgert, R. (1974): *Caiman crocodilus apaporiensis*, Medem, 1955, im Terrarium.—DATZ, 27: 356–358.

Hirschfeld, K. (1966): Paarung und Eiablage der Brillenkaimane im Vivarium Kehl.—DATZ, 19: 151–154.

—— (1966): Zucht von Krokodilkaimanen (*Caiman crocodilus*) im Vivarium Kehl.—DATZ, 19: 308–310.

—— (1967): Der Kaiman-Nachwuchs im Vivarium Kehl.—DATZ, 20: 217–219.

—— (1969): Die Geburt eines Krokodil-Kaimans.—Aquarien-Mag., 2: 52–55.

—— (1976): Die letzten ihrer Art.—Das Terrarium, 2: 28–31.

—— (1976): Australische Stachelskinke, Zucht im Vivarium Kehl.—Das Terrarium, 2: 36–37.

Hitz, R. (1983): Pflege und Nachzuchten von *Trachydosaurus rugosus*, Gray, 1827, im Terrarium (Sauria: Scincidae).—Salamandra, 19 (4): 198–210.

Hoesche, U. (1985): *Zonosaurus laticaudatus* (Grandidier).—Sauria, 7 (3): 2.

Hofer, R. (1977): Fortpflanzung und Aufzucht mediterraner Eidechsen im Terrarium.—Das Aquarium 92: 77–82.

Honegger, R. E. (1975): Beitrag zur Kenntnis des Wickleskinkes *Corucia zebrata*.—Salamandra, 11 (1): 27–32.

—— & C. R. Schmidt (1964): Herpetologisches aus dem Züricher Zoo. Beiträge zur Haltung und Zucht verschiedener Reptilien.—DATZ, 17: 339–342.

Horn, H. G. (1977): Notizen zu Systematik, Fundortangaben und Haltung von *Varanus* (*Varanus*) *karlschmidti* (Reptilia: Sauria: Varanidae).—Salamandra, 13 (2): 78–88.

—— (1978): Nachzucht von *Varanus gilleni* (Reptilia: Sauria: Varanidae).—Salamandra, 14 (1): 29–32.

—— (1980): Bisher unbekannte Details zur Kenntnis von *Varanus varius* auf Grund von feldherpetologischen und terraristischen Beobachtungen (Reptilia: Sauria: Varanidae).—Salamandra, 16 (1): 1–18.

—— & G. Peters (1982): Beiträge zur Biologie des Rauhnackenwarans, *Varanus* (*Dendrovaranus*) *rudicollis*, Gray (Reptilia: Sauria: Varanidae).—Salamandra, 18 (1/2): 29–40.

—— & B. Schulz (1977): *Varanus dumerilii* wie ihn nicht jeder kennt.—Das Aquarium 91: 37–38.

—— & U. Schürer (1978): Bemerkungen zu *Varanus* (Odatia) *glebopalma*, Mitchell, 1955 (Reptilia: Sauria: Varanidae).—Salamandra, 14 (3): 105–116.

Houba, J. (1957): *Chalcides ocellatus ocellatus*, Forskal, die Walzenechse.—DATZ, 10: 192–194.

Houtman, H. (1988): De verzorging en voortplanting van Gerrards blauwtongskink (*Tiliqua gerrardii*).—Lacerta, Den Haag, 46 (6): 92–95.

Inger, R. F. & B. Greenberg (1966): Annual reproductive patterns of lizards from a Bornean rainforest.—Ecology, 47 (6): 1006-1021.
Irwin, B. (1986): Captive breeding of two species of monitor.—thylacinus, 11 (2): 4-5.
Jauch, D. (1984): Vermutlich eine Weltpremiere: Die Nachzucht des Stachel-schwanz-Warans.—Aquarien-Mag., 16: 236-237.
Jelden, D. & H. Frey (1989): Die größten Krokodilfarmen der Erde.—DATZ, 42: 353-357.
Jensen, J. K. (1982): Relations betwen temperature and incubation time for eggs of the sand lizard (*Lacerta agilis* L.). Amphibis-Reptilia. Wiesbaden, 2: 385-386.
Jes, H. (1955): Beobachtungen an jungen *Caiman c. crocodylus* (L.) und *Paleosuchus trigonatus* (Schn.)—DATZ, 8: 161-162.
John, W. (1980: Beobachtungen im Biotop der Braunen oder Sudan-Schildechse (*Gerrhosaurus major*) A. Duméril, 1851.—herpetofauna, 5: 31-32.
Jungnickel, J., R. Sommerlad & L. Trutnau (1991): Einige biologische Daten zur Kenntnis des Leistenkrokodils *Crocodylus porosus*, Schneider, 1801.—Herpetofauna 13 (71): 27-34.
Kabisch, K. (1986): *Podarcis taurica* (Pallas, 1814)—Taurische Eidechse.—In: Böhme, W. (Hrsg.): Handbuch der Reptilien und Amphibien Europas: 343-362.
Katzenstein, H. (1967): *Eumeces schneideri algeriensis*, der Berberskink.—DATZ, 20: 185-186.
Kenneweg, F. H. (1956): Meine Beobachtungen an der Netzwühle, *Blanus cinereus* Vand.—DATZ, 9: 77-78.
Klag, K. & H. Krantz, (1988): Bemerkungen zur Haltung und Fortpflanzung von *Varanus b. bengalensis* im Terrarium.—herpetofauna, 10 (52): 21-24.
Kober, J. (1990): Im Terrarium gezüchtet: *Gerrhosaurus flavigularis*.—DATZ, 43: 24-25.
Koore, J., van de (1988/89): De kweek van varanen in gevangenshap.—Lacerta, Den Haag, 47 (2): 35-39.
Kopstein, F. (1932): Herpetologische Notizen VI. Weitere Beobachtungen über die Fortpflanzung westjavanischer Reptilien.—Treubia, Bogor, 14 (1): 78-83.
—— (1938): Ein Beitrag zur Eierkunde und Fortpflanzung der malaiischen Reptilien.—Bull. Raffles Mus., Singapore, 14, 81-167, 22 pls., 75 figs.
Köhler, G. (1989): *Tupinambis teguixin* (Linnaeus).—Amph./Rep.-Kartei, Beilage in Sauria, Berlin-W., 11 (1): 133-136.
Krabbe-Paulduro, U. (1987): *Cordylus jonesii* (Boulenger).-Amph./Rept.-Kartei, Beilage in Sauria, Berlin-W., 9 (4): 93-96.
—— & E. Paulduro (1989): *Cordylus warreni* (Boulenger).— Amph./Rept.-Kartei, Beil. Sauria, 11 (4): 153-160.
Kratzer, H. (1973): Beobachtungen über die Zeitgungsdauer eines Eigeleges von *Varanus salvator* (Sauria, Varanidae).—Salamandra, 9 (1): 27-33.
Krebs, U. (1979): Der Duméril-Waran (*Varanus dumerilii*), ein spezialisierter Krabbenfresser? (Reptilia: Sauria: Varanidae).—Salamandra, 15 (3): 146-157.
Krintler, K. (1986): Dominica-herpetologisches Kleinod in der Karibik.—herpetofauna, 8 (45): 26-30.
Kuchling, G. (1970: Beobachtungen an *Eumeces schneideri*.—DATZ, 23: 216-217.
Kutschera, U. (1976): Aufzucht und Pflege des Zwerg-Gürtelschweifes, *Cordylus cordylus*.—DATZ, 29: 282-283.
Ladiges, W. (1939): Herpetologische Beobachtungen auf Sumatra.—Zool. Anz., Leipzig, 128: 235-249.

Langerwerf, B. (1979): De Alligatorhagedis, *Gerrhonotus multicarinatus*, als bewoner van een buiteterrarium.—Lacerta, Den Haag, 37 (7): 107-111.

— (1980): The Caucasian green lizard, *Lacerta strigata*, Eichwald, 1831, with notes on its reproduction in capitivity.—Brit. Herp. Soc. Bull., 1: 23-26.

— (1981): Nigrinos bei der Nachzucht von *Lacerta lepida pater*.—herpetofauna, 12: 21-22.

— (1984): Verzorging en kweek van de Oman hagedis, *Lacerta jayakari*.—Lacerta, Den Haag, 43 (1): 4-6.

Laube, A. (1981): Freiland-und Gefangenschaftsbeobachtungen am Algerischen Sandläufer—*Psammodromus aligiris algirus* (Linnaeus).—Sauria, 4 (3): 7-10.

Laue, H. (1982): Erfahrungen bei der Haltung von *Varanus bengalensis*.—Elaphe, 4 (1): 1-4.

Laurens, B. L. (1976): Erfahrungen mit dem Blauschwanzskink, *Mabuya quinquetaeniata*.—DATZ, 29: 28-31.

Lehmann, H. (1987): Hypothetische Überlegungen zur Schlupfproblematik von künstlich inkubierten Gelegen südamerikanischer Schildkrötenarten der Familie Chelidae.—Salamandra, 23 (2/3): 73-77.

Liesack, H. (1984): Wüstenrenner aus Transkaukasien.—Elaphe, 6 (1): 2-4.

— (1988): Erfahrungen bei der Haltung von *Lacerta agilis brevicaudata*.—Elaphe, 10: 25-27.

Lim Boo Liat (1976): Lizards Galore.—Nat. Malays., 1 (2): 12-19, 16 figs.

Lüthi, H. (1983): Haltung und Zucht des Brauen-Glattstirnkaimans (*Paleosuchus palpebrosus*).—herpetofauna 24: 22-26.

— (1986): Haltung und Zucht des Brauen-Glattstirnkaimans (*Paleosuchus palpebrosus*).—Elaphe (Sonderhelft): 98-100.

— & P. H. Stettler (1977): Panzerechsen als „Hausgenossen"? *Paleosuchus palpebrosus* im Zimmerterrarium.—Das Aquarium 91: 31-36.

Mägdefrau, H. (1987): Zur Situation der Chinesischen Krokodilschwanz-Höckerechse, *Shinisaurus crocodilurus*, Ahl, 1930.—herpetofauna, 9 (51): 6-11.

Malkmus, R. (1983): Herpetologische Exkursionen im Koken Atlas/Marokko.—herpetofauna, 27: 8-12.

Mantel, P. (1986): Terrariumervaringen met de slangooghegedes, *Ophisops elegans*.—Lacerta, Den Haag, 44 (9): 154-158.

Manthey, U. (1981): Die Echsen des ceylonesischen Regenwaldes und seiner Randgebiete.—Sauria, 3 (2): 25-35.

— (1983): Verhaltensfehldeutung von *Dasis smaragdina* infolge falscher Terrarienabmessungen.—Sauria, 5 (3): 23-24.

— (1985): Der Smaragdskink, *Lamprolepis smaragdina* (Lesson). Ein eierfressender Baumskink.—Sauria, 7 (1): 5-11.

— & W. Denzer (1982): Exkursion am Mt. Kinabalu 4100 m, Nordborneo. Teil 2: Herpetologische Eindrücke.—herpetofauna, 4 (21): 11-19.

Matz, G. (1971): *Mabuya capensis* Gray.—DATZ, 24: 96-97.

— (1972): Einige australische Skinke der Gattungen *Egernia* und *Tiliqua*.—DATZ, 25: 136-139.

— (1973): Gürtelechsen.—DATZ, 26: 28-29.

— (1973): Gürtelechsen II.—DATZ, 26: 65-67.

— (1975): *Platysaurus guttatus*.—DATZ, 28: 361-362.

— (1977): Die Tejus der Gattung *Tupinambis*.—DATZ, 30: 172-174.

Mebs, D. (1974): Haltungserfahrungen mit *Tiliqua casuarinae* (Sauria: Scincidae).—Salamandra 10, (3/4): 106–106.
Meier, H. (1979): Herpetologische Beobachtungen auf Neukaledonien.—Salamandra, 15, (3): 113–139.
—— (1988): Zur Ökologie, Ethologie und Taxonomie einiger Schildechsen der Gattung *Tracheloptychus* und *Zonosauru* auf Madagaskar, Teil 1.—herpetofauna, 10 (57): 22–26.
—— (1989): Zur Ökologie, Ethologie und Taxonomie einiger Schildechsen der Gattungen *Tracheloptychus* und *Zonosaurus* auf Madaskar, Teil 2.—herpetofauna, 11 (58): 14–23.
Mertens, R. (1939): Herpetologische Ergebnisse einer Reise nach der Insel Hispaniola, Westindien.—Abh. senkenberb. naturf. Ges., 449: 1–84.
—— (1964): Beobachtungen an Taubwaranen (*Lanthanotus borneensis*) im Terrarium.—DATZ, 17: 179–183.
—— (1966): Nachtrag zu Beobachtungen an Flossenfußechsen (pygopodidae).—DATZ, 19: 96.
—— (1969): Zur Haltung von Blindwühlen (*Gymnophona*).—DATZ, 22: 81–82.
Mudrack, W. (1974): Der Rosazungenskink—eine terraristische Kostbarkeit.—Aquarien-Mag., 8: 407–411.
—— (1976): Ein Leben unter Tage. Der Prachtskink, seine Pflege und Zucht.—Aquarien-Mag., 10: 120–121.
—— (1977): Wenn's kühl und feucht wird, taut er auf. *Cophoscincopus durus*, ein kleiner Wasserskink.—Aquarien-Mag., 11: 282–283.
—— (1985): Ein Beitrag zur Haltung und Zucht des Wasserskinkes *Cophoscincopus durus* (Cope, 1862).—Sauria, 7 (1): 29–30.
Münsch, W. (1986): Trinkwasser für Stutzechsen.—Das Aquarium, 207: 496.
Nettmann, H.-K. & S. Rykena (1984): *Lacerta trilineata*, Bedriaga, 1886—Riesensmaragdeidechse.—In: Böhme, W. (Hrsg.): Handbuch der Reptilien und Amphibien Europas: 100–128.
—— (1984): *Lacerta viridis* (Laurenti, 1768)—Smaragdeidechse.—In: Böhme, W. (Hrsg.): Handbuch der Reptilien und Amphibien Europas: 129–179.
Niekisch, M. (1975): Pflege und Nachzucht von *Ergenia cunninghami* (Sauria: Scincidae).—Salamandra, 11 (3/4): 130–135.
—— (1980): Terraristische Beobachtungen zur Biologie von *Egernia cunninghami* (Reptilia: Sauria: Scincidae).—Salamandra, 16 (3): 162–176.
—— (1981): *Chamaesaura anguina* im Terrarium.—herpetofaun, 13: 9–10.
Nöllert, A. (1983): Einige Bemerkungen zur Taurischen Eidechse, *Podarcis taurica taurica* (Pallas), in Südostbulgarien.—herpetofauna, 25: 26–29.
Obst, F. J., K. Richter & U. Jacob (1984): Lexikon der Terraristik und Herpetologie.—Landbuch-Verlag. Hannover.
Osenegg, K. (1989): Die Amphibien und Reptilien der Insel Zypern.—Diplomarbeit der Universität Bonn, 200 S.
—— (1991): Forschungsprojekt: Evolution der Viviparie bei *Lacerta vivipara*.—Die Eidechse, 2: 10–13.
Ott, M. (1986): Ruineneidechsen im Terrarium gepflegt und gezüchtet.—Das Aquarium, 201: 145–149.
Pambov, M. (1985): Fang, Haltung und Zucht von *Lacerta trilineata dobrogica*.—Elaphe, 7 (1): 8–9.

Paulduro, E. & U. Krabbe-Paulduro (1987): Anmerkungen zur Lebensweise, Haltung und Zuchtproblematik des Risengürtelschweifs *Cordylus giganteus*, A. Smith, 1844.—Sauria, 9 (4): 9-17.

Peters, G. (1989): Echsen, ein Brevier. Edition Leipzig.

—— & R. Schubert (1968): Theoretische und praktische Aspekte einer biologischen Expedition durch Kub.—Biol. id. Schule, 8/9: 339-346.

Peters, J. H. & R. Donoso-Barros (1970): Catalogue of the Neotropical Squamata, Part II. Lizards and Amphisbaenians.—U.S. nation, Mus. Bull., 297: I-VII, 1-293, 104 figs.

Peters, U. (1965): Beobachtungen an *Xantusia riversiana*.—DATZ, 18: 52-54.

—— (1967): Beobachtungen an *Varanus varius* (Shaw) in der Natur.—DATZ, 20: 120-121.

—— (1968): *Moloch horridus*, *Varanus spenceri*, *V. mitchelli*, *Egernia bungana* und *Heteronotia binoei* im Taronga-Zoo, Sydne.—DATZ, 21: 252-254.

—— (1969): Zum ersten Mal in Gefangenschaft: Eiablage und Schlupf von *Varanus spenceri*.—Aqu. Terr., 16 (9): 306-307.

—— (1974): Kurz vorgestellt: *Varanus* (Odatria) *gilleni*.—Das Aquarium 65: 512-513.

—— (1975): Das Echsenporträt. Der Smaragd-Skink.—Aquarien-Mag., 9: 345.

—— (1986): Gelungene Aufzucht von *Varanus spenceri* (Lucas & Frost).—Das Aquarium, 205: 377-379.

Peters, U. W. (1973): Ein Beitrag zur Ökologie von *Varanus* (Odatria) *storri* Mertens, 1966.—Das Aquarium, 53: 462-463.

—— (1985): Bemerkenswerte Reptilien aus der pazifischen Inselwelt.—Das Aquarium, 191: 254-261.

Petzold, H.-G. (1965): Über Freileben und Terrarienhaltung der Taurischen Eidechse.—Aqu. Terr., 12: 112-117.

—— (1969): *Cricosaura typica*, Gundlach & Peters, eine herpetologische Kostbarkeit aus Kuba.—DATZ, 22: 82-85.

—— (1981): *Mabuya aurata septemtaeniata* (Reuss, 1834), Östlicher Goldskink (Familie: Scincidae, Skinke).—Aqu. Terr., 28 (8): 288.

Polleck, R. (1980): Ein temperamentvoller Zwerg im Terrarium: Der Gillenwaran (*Varanus gilleni*) Lucas & Frost, 1895.—herptofauna, 6: 19-20.

—— (1982): Ein seltener Terrariengast: Der Rostkopfwaran (*Varanus semiremex*, Peters, 1869.—herpetofauna, 16: 19-20.

Reiß, G. (1961): Aufzucht von Ruineneidechsen.—DATZ, 14: 212-214.

Rese, R. (1981): Schildechse (Gerrhosaurus)—Haltung und Zucht.—Sauria, 3 (3): 29-31.

—— (1983a): Der Timorwaran *Varanus timorensis*—Haltung und Zucht.—Sauria, 5 (4): 13-15.

—— (1983b): Seltene Nachzucht im Terrarium—Die Zucht des Steppenwarans—*Varanus exanthematicus*.—Sauria, 5 (3): 25-28.

—— (1984): Heute schon eine Kostbarkeit: Der Duméril-Waran, *Varanus dumerilii* (Schlegel, 1839).—Sauria, 6 (2): 25-27.

—— (1984): Der Zwergwaran, *Varanus storri*, Mertens, 1966.—Sauria 6 (1): 33-34.

—— (1986): *Varanus dumerilii* (Schlegel).—Sauria, 8 (1): 2.

—— (1986): *Varanus storri*, Mertens.—Amph./Rept.-Kartei, 55-56—Beilage in Sauria, Berlin-W., 8 (3).

—— (1986): *Gerrhosaurus major*, Duméril.—Amph./Rept.-Kartei, 61–62—Beilage in Sauria, Berlin-W., 8 (4).
Richter, K. (1986a): *Podarcis dugesii* (Milne-Edwards, 1829)—Madeira-Mauereidechse.—In: Böhme, W. (Hrsg.): Handbuch der Reptilien und amphibien Europas: 388–398.
—— (1986b): *Podarcis perspicillata* (Duméril Bibron, 1839)—Brilleneidechse.—In: Böhme, W. (Hrsg.): Handbuch der Reptilien und Amphibien Europas: 399–407.
Richter, U. (1981): Eine Nachzucht des Stumpfkrokodils (*Osteolaemus tetraspis*) (Reptilia: Crocodylia: Crocodylidae).—Salamandra, 17 (3/4): 194–197.
Rieppel, O. (1973): Lithopagie bei Reptilein.—Das Aquarium, 49: 292–293.
Rinke, D. (1986): Herpetologisches aus Fidschi und Tonga.—DATZ, 39: 226–230.
Riosuke, A. (1985): Beobachtungen an *Crocodylus mindorensis* auf Mindoro, Philippinische Inseln.—herpetofauna, 7 (34): 6–19.
Rödel, M. O., R. Bussmann, & A. Kaupp (1989): Beitrag zur Biotopwahl von *Ophiomorus punctatissimus* (Bibron & Bory 1833).—Sauria, 11, (1): 27–30.
Rogner, M. (1979): Exkrusion zu den Kanarischen Inseln.—Aquarien-Mag., 13: 456–461.
—— (1982): Durch Nachzucht erhalten: Der gestreifte Kanarenskink.—Aquarien-Mag., 16: 284–285.
—— (1983): Zur Biologie, Pflege und Zucht einiger Walzenskink-Arten *(Chalcides)*.—Das Aquarium, 174: 657–662.
—— (1984): Ein grüner Goliath: Die Riesensmaragdeidechse. Zur Pflege und Zucht von *Lacerta trilineata*.—Aquarien-Mag., 16: 176–178.
—— (1987): *Lacerta oxycephala*. Beobachtungen im Lebensraum der Spitzkopfeidechse.—Das Aquarium, 212: 93–96.
—— (1991): Grüne Madagaskar-Schildechse—Anmerkungen zur Haltung und Zucht von *Zonosaurus haraldmeieri*, Brygoo & Böhme, 1985.—Das Aquarium, 259: 47–50.
—— & H. D. Philippen (1985): Tyrrhenische Gebirgseidechse. Zur Biologie und Ökologie von *Lacerta bedriagae bedriagae* Camerano, 1885.—Das Aquarium, 198: 651–656.
Rohde, H. (1989): Zur Pflege von *Enyaliosaurus quinquecarinatus* (Gray, 1842).—Sauria, 11 (2): 11–13.
Rooij, N. de (1915): The reptiles of the Indio-Australian Archepelago. I. Lacertilia, Chelonia, Emydosauria.—Leiden (E. J. Brill): xiv + 384 pp., 132 figs.
—— (1922): Fauna Simaluensis Reptilia.—Zool. Meded., Leiden, 6: 217–238, 8 figs.
Rössler, R. (1992): Die grüne Langschwanzeidechse—*Takydromus smaragdinus*. Ein Haltungs-und Zuchtbericht.—Die Eidechse, 5: 3–5.
Röstel, G. S. (1987): Legenot bei einer *Lacerta lepida*, Daudin, 1802, und einem *Chamaeleo chamaeleon* (Linnaeus, 1758)—Eine Methode zur Austreibung von Resteiern.—Salamandra, 23 (2/3): 179–180.
Rudloff, H. W. (1986): *Paleosuchus palpebrosus* Brauen-Glattstirnkaiman.—Elaphe, 1: 20.
Rust, R. (1957): Seltene Gäste aus der Gyla-Wüste.—DATZ, 10: 81–82.
Rüegg, R. (1974): Nachzucht beim Timor-Baumwaran, *Varanus timorensis similis*, Mertens, 1958).—Das Aquarium, 62: 360–363.
Rutschke, J. (1989): Erfahrungen bei langjähriger Haltung und Zucht der Perleidechse, *Lacerta lepida leida*, Daudin, 1802, unter besonderer Berücksichtigung des UV-Einflusses auf die Vitalität der Jungtiere.—herpetofauna, 11 (60): 25–31.

Rykena, S. & L. Henke (1978): Bastardierung von *Lacerta viridis* und *Lacerta agilis* im Terrarium (Reptilia: Sauria: Lacertidae).—Salamandra, 14 (3): 147-152.

—— & H. K. Nettmann, & W. Bings (1977): Zur Biologie der Zagros-Eidechse, *Lacerta principes*, Blanford, 1874, I. Beobachtungen im Freiland und im Terrarium an *Lacerta princeps kurdistanica*, Suchov, 1936 (Reptilia: Sauria: Lacertidae).—Salamandra, 13 (3/4): 174-184.

—— (1978): Die Iberische Smaragdeidechse. Aquarien.-Mag., 1978): 56-58.

Salvador, A. (1984): *Lacerta schreiberi*, Bedriaga, 1878—Iberische Smaragdeidechse.—In: Böhme, W. (Hrsg.): Handbuch der Reptilien und Amphibien Europas: 69-81.

—— (1984): *Lacerta monticola*, Boulenger, 1905—Iberische Gebirgseidechse.—In: Böhme, W. (Hrsg.): Handbuch der Reptilein und Amphibien Europas: 276-289.

—— (1986): *Podarcis hispanica* (Steindachner, 1870):-Iberische Mauereidechse.—In: Böhme, W. (Hrsg.): Handbuch der Reptilien und amphibien Europas: 71-82.

—— (1986): *Podarcis lilfordi* (Günther, 1874)—Balearen-Eidechse.—In: Böhme, W. (Hrsg.): Handbuch der Reptilien und Amphibien Europas: 83-110.

—— (1986): *Podarcis pityuensis* (Bosca, 1883)—Pityusen-Eidechse.—In: Böhme, W. (Hrsg.): Handbuch der Reptilien und Amphibien Europas: 231-253.

Sauer, F. (1989): Kriechtiere und Lurche im Mittelmeerraum, nach Farbfotos erkannt.—Fauna-Verlag, Karlsfeld.

Sauterau., L. & P. de Bitter (1980): Notes sur l'élevage et la reproduction en captivité du varan de timor (Sauria: Varanidae).—Bull. Soc. herpetol. France: 4-9.

Schade, W. (1977): Meine Blauzungenskinke, *Tiliqua scincoides*.—Das Aquarium, 99: 389-390.

—— (1980): *Tiliqua gigas* (Risenskink). Meine ersten Erfahrungen mit Pflege und Nachzucht.—Sauria, 2 (3): 23-24.

Schneider, B. (1986): *Podarcis tiliguerta* (Gmelin, 1789)—Tyrrhenische Mauereidechse.—In.: Böhme, W. (Hrsg.): Hb. d. Reptilien und Amphibien Europas: 363-376.

Schmida, G. E. (1971): Der getüpfelte Baumwaran Australiens, *Varanus timorensis similis*.—DATZ, 24: 168-170.

—— (1974): Der Kurzschwanzwaran (*Varanus brevicauda*).—DATZ, 27: 390-394.

Schmidt, A. A. (1991): Zur Haltung und Fortpflanzung des Wickelskinks, *Corucia zebrata*, Gray, 1855.—Salamandra, 27 (4): 238-245.

Schmidt, D. (1981): Echsen im Terrarium.—Vivaristik Ratgeber 2, Neumann-Verlag. Leipzig, Radebeul.

—— (1986): *Mabuya perrotetii*, Togoskink.—Elaphe, 3: 60.

—— (1988): *Mabuya septemtaeniata* Goldmabuye.—Elaphe, 10:40.

Schmidtler, J. F. (1986): Orientalische Smaragdeidechsen: 2. Über Systematik und Synökologie von *Lacerta trilineata*, *L. media* und *L. pamphylica* (Sauria: Lacertidae).—Salamandra, 22 (2/3): 126-146.

Schönfeld, W. (1973): Geburt und Aufzucht von *Cordylus cataphractus*.—DATZ, 26: 390-391.

Schröder, J. H (1969): Biologie und Haltung der kaukasischen Felseidechse, *Lacerta saxicola*, Eversmann,—DATZ, 22: 52-53.

Schulze, E. (1971): Riesenechsen *Tupinambis nigropunctatus*.—Das Aquarium, 28: 595.

Schürer, U. & H. G. Horn (1976): Freiland und Gefangenschaftsbeobachtungen am australischen Wasserwaran, *Varanus mertensi*.—Salamandra, 12 (4): 176-188.

Schwartz, A. & R. W. Henderson (1985): A Guide to the Identification of the Amphibians and Reptiles of the West Indies. Exclusive of Hispaniola.—Milwaukee public Mus.: 1-165.
— (1988): West Indian Amphibians and Reptiles: A Check-List.—Contributions Biol. Geol. Nu. 74, Milwaukee, public Mus.: 1-264.
Shaw, Ch, E. (1968): Reproduction of the Gila Monster (*Heloderma suspectum*) at the San Diego Zoo.—Zool. Garten (N.F.), 35 (1/2): 1-6.
Smith, M. A. (1930): The Reptilia and Amphibia of the Malay Peninsula.—Bull. Raffles Mus., Singapore, 3: 1-149.
— (1935): The fauna of British India, Ceylon and Burma. Reptilia and Amphibia. Voll. II. Sauria. London (Taylor & Francis): I-xiii, 1-440; 1 pl, 94 figs.
— & K. G. Gardner (1915): List of mammals, birds, reptiles and batrachians obtained in the Ratburi and Petchaburi districts.—J. nat. Hist. Soc. Siam, London, 1 (3): 146-156.
Sprackland, R. G. (1989): The Chinese Crocodile Lizard.—Tropical Fish Hobbyist, 38: 102-109.
Stemmler-Morath, C. (1958): Einige Beobachtungen an der Brückenechse.—DATZ, 11: 151-152.
Stirnberg, E. & H. G. Horn (1981): Eine unerwartete Nachzucht im Terrarium: *Varanus* (Odatria) *storri* (Reptilia: Sauria: Varanidae).—Salamandra, 17 (1/2): 55-62.
Stribrny, R. (1978): Nachzucht von *Caiman crocodilus fuscus* in der Gefangenschaft.—DATZ, 31: 422-424.
Stuppi, M. (1984): Zum Aggressionsverhalten von Krokodilen.—DATZ, 37: 152-154.
Switak, K. H. (1978): Leben in der Wüste. Die Nachtechsen *Xantusia vigilis* und *Xantusia henshawi*.—Das Aquarium, 107: 224-227.
— (1979): Die Felsen-Schildechse *Gerrhosaurus v. validus*.—Aquarien-Mag., 13: 144-145.
— (1980): Der Riesengütelschweif oder Südafrikanische Sonnengucker, *Cordylus gigatneus*.—Das Aquarium, 136: 537-541.
— (1980): Vom Sonnenanbeter und anderen Gürtelschweifen.—Aquarien-Mag., 14: 317-321.
— (1981): Leben in der Wüste: Die Westliche Rennechse, *Cnemidophorus tigris*.—Das Aquarium, 156: 321-323.
— (1987): Der Panzergürtelschweif *Cordylus cataphractus*, Boie, 1828.—Das Aquarium, 221: 589-591.
— (1988): Aus dem Leben der Felsenschildechse *Gerrhosaurus validus*.—Das Aquarium, 223: 45-46.
Tannher, V. M. (1953): Pacific island herpetology no. VII,—Ulu Langat, State of Selangor, Malay. Great Basin Natural. Provo, 13 (1-2): 1-7.
Taylor, E. H. (1953): A review of the lizards of Ceylon. Kansas Univ. Sci. Bull., Lawrence, 44: 647-1077.
Thieme, U. (1983): *Eumeces schneideri princips* und *Mabuya aurata septemtaeniata*—zwei selten gepflegte Glattechsen aus Transkaukasien.—herpetofauna, 22 (Febr.): 33-34.
Tiedemann, F. & K. Henle (1986): *Podarcis melisellensis* (Braun, 1877)—Adriatische Mauereidechse, Karstläufer.—In: Böhme, W. (Hrsg.): Handbuch der Reptilien und Amphibien Europas: 111-141.
Trautmann (1924): In: Vereinsnachrichten Vivarium Heidelberb. B. Aqu. Terr.kde., 35: 253-254.

Trutnau, L. (1968): Gefangenschafts-Beobachtungen an Krustenechsen (*Heloderma suspectum*, Cope).—DATZ, 21: 120-125.
—— (1976): Zum Fortpflanzungsverhalten der Skorpions-Krustenechse (*Heloderma h. horridum*).—DATZ, 29: 424-428.
—— (1984): Herpetofocus, das Gilatier.—herpetofauna, 32 (Okt.): 13.
—— (1990): Zur Kenntnis des Krokodilkaimans, *Caiman crocodilus* (Linnaeus, 1758).—herpetofauna, 12 (64): 25-34.
Ulrich, D. (1989): Die Everglades: Alligatoren, Krokodile und Kaimane.—DATZ, 42: 685-688.
Verbeek, B. (1972): Über Haltung und Zucht von *Lacerta hispanica* (Sauria: Lacertidae).— Salamandra, 8 (3/4): 183-185.
Villa, J., L. D. Wilson & J. D. Johnson (1988): Middle American herpetology: A Bibliographic Checklist.—Univ. Missouri Press, Columbia, 132 S.
Visser, G. (1989): Chinese krokodilstaart hagedissen (*Shinisaurus crocodilurus*) in Diergaarde Blijdorp.—Lacerta, Den Haag, 47 (4): 98-105.
—— (1985): Notizen zur Brutbiologie des Gelbwarans, *Varanus* (Empagusia) *flavescens* (Hardwicke & Gray, 1827), im Zoo von Rotterdam (Sauria: Varanidae).—Salamandra, 21 (2/3): 161-168.
Vogel, Z. (1962): Die Raubechsen Südamerikas.—DATZ, 15: 372-376.
Voris, H. U. (1977): Comparison of Herpetofauna diversity in tree Buttresses of evergreen tropical forests.—Herpetologica, 33: 375-380.
Wanrooy, V. v. (1987): De zes streep-lag-staat-hagedis (*Takydromus sexlineatus*) in het terrarium.—Lacerta, Den Haag, 45, (7): 102-106.
Welsch, H. (1985): Verhaltensbeobachtungen an Nilkrokodilen, *Crocodylus niloticus*, Laurenti, 1768, auf einer Farm (Crocodylia: Crocodylidae).—Salamandra, 21 (2/3): 104-122.
Wermuth, H. (1953): Systematik der rezenten Krokodile.—Mitt. zool. Mus. Berlin, 29 (1): 375-514.
—— (1956): Internationaler Naturschutz für Krokodile!—Aqu. Terr., Berlin/Leipzig, 3: 248-250.
—— (1963): Farbwechsel und Lernfähigkeit bei Krokodilen.—DATZ, 16 (3): 90-92.
—— (1964): Das Verhältnis zwischen Kopf-, Rumpf- und Schwanzlänge bei den rezenten Krokodilen.—Senckenb. biol., 45 (7-5): 309-385.
—— (1965): Liste der rezenten Amphibien und Reptilien. Gekkonidae Pygopodidae, Xantusiidae.—Das Tierreich.—Berlin, 80; I-XXII, 1-246.
—— (1967): Liste der rezenten Amphibien und Reptilien; Agamidae.—Das Tierreich, Berlin, 86; I-xiv, 1-127.
—— (1972): Krokodilschutz und Krokodilleder-Industrie.—DATZ, 25 (7): 249-251.
—— (1975): Wohin mit „überflüssigen" Schildkröten und Krokodilen?—DATZ, 28 (2): 104-105.
—— (1979): Krokodile müssen überleben. Industrie fordert Nachzucht.—Sielmanns Tierwelt, 3 (5): 4.
—— & K. H. Fuchs (1978): Bastarde zwischen südostasiatischen Krokodilen.—Stuttgart. Beitr. Naturk., (A) 314: 1-17.
—— & K. H. Fuchs (1983): Order Crocodylia.—Identification Manual, 3 (Reptilia, Amphibia, Pisces), Convent. Internation. Trade in Endangered Species (Cites).

—— & Mertens, R. (1977): Liste der rezenten Amphibien und Reptilien: Testudines, Crocodylia, Rhynchocephalia.—Das Tierreich, Berlin, 100: I–XXVII, 1–174.
Werner, F. (1900): Reptilien und Batrachier aus Sumatra.—Zool. Jb. (Syst.), Jena, 13: 479–508; 5 Abb.
Whitaker, R. & Z. Whitaker (1983): Arterhaltung durch Nachzucht von Krokodilen in Indien.—DATZ, 36: 220–224.
Wilke, H. (1985): Krokodilschwanz-Höckerechsen im Vivarium Darmstadt.—DATZ, 38: 234.
—— (1985): Erfolgreiche Eingewöhnung von seltenen Krokodilschwanz-Höckerechsen im Vivarium, Darmstadts Tiergarten, gelungen.—herpetofauna, 7 (34): 30.
—— (1985): Eingewöhnung, Haltung und Geburt der seltenen Krokodilschwanz-Höckerechsen, *Schinisaurus crocodilurus*, Ahl, 1930.—Sauria, 7 (1): 3–4.
Wittig, W. (1980): Küunstliche Erbrütung der Eier von Lacerten.—Elaphe, 2 (1): 3–5.
—— (1986): Zur Vermehrung der Smaragdeidechse und verwandter Arten.—Elaphe, 8 (4): 61–65.
—— (1987): Zur Zucht einiger Arten Mauer- und Felseneidechsen.—Elaphe, 9 (3): 42–44.
Zimmermann, H. & E. Zimmermann (1983): Durch Nachzucht erhalten: Die Fransenfinger-Eidechse *Acanthodactylus boskianus*.—Aquarien-Mag., 17: 386–390.
Zobel, R. (1988): Eine herpetologische Expedition durch Madagaskar 2. Teil, DATZ, 4: 236–239.
Zwinenberg, A. J. (1974): Die Smaragdeidechse *Lacerta virdis* im Terrarium.—Das Aquarium, 56: 78–79.

Photographic Sources

Baur, Koppingen (Switzerland)	pp 40, 100
Bischoff, Bonn (Germany)	p 166 (below)
Böhme, Bonn (Germany)	pp 58 (above), 60 (above), 166 (above)
Dell'mour, Tullnerbach (Austria)	pp 47, 67 (below), 109 (below), 137, 138, 140, 155, 205, 222 (below)
Ebbert, Bruch (Germany)	pp 174, 175
Eidenmüller, Frankfurt (Germany)	pp 22, 271 (below)
Franzen, Bonn (Germany)	pp 90, 101, 104 (below), 108 (above and below), 115 (above and below), 125, 127, 218, 246
Glaw/Müller, Cologne (Germany)	pp 56, 60 (below)
Glaw/Vences, Cologne (Germany)	pp 62 (above), 65 (above)
Grossmann, Berlin (Germany)	pp 171, 172
König, Kiel (Germany)	pp 4, 24, 28, 34, 77, 163, 185 (below) 196, 230, 231, 236, 250, 258, 260, 264, 282 (above and below)
Langerwerf, Monte Vallo (U.S.A.)	pp 42 (left), 128
Mägdefrau, Lauf (Germany)	p 2
Manthey, Berlin (Germany)	pp 15, 209 (above and below), 215 (above and below)
Maronde, Berlin (Germany)	p 42 (right)
Mudrack, Berlin (Germany)	pp 206 (below) 219, 226, 237
Natur & Text, Kühnel & Schwarzer, Berlin, (Germany)	pp 133, 158, 190 (left), 262, 275 (above)
Pflugmacher, Augsburg (Germany)	p 239
Polak, Raknovik (Czech Republic)	p 243
Rogner, Hürtgenwald (Germany)	pp 11, 30 (left), 44, 50, 58 (above), 62 (below), 64, 95, 104 (above), 119 (above), 135, 161, 190 (right), 192, 203, 222 (above) 271 (above and below), 243
Sauer, Karlsfeld (Germany)	pp 27, 87, 143, 144 (above and below), 162
Switak, Santa Rosa (U.S.A.)	pp 30 (right), 39, 52, 69, 72 (above and below), 74, 81, 179 (above and below), 183, 185 (above), 186, 206 (above), 212, 228

Index

Ablepharus, 202–203
—*kitaibeli*, 202–203
Acanthodactylus, 83–86
—*boskianus*, 84
—*erythrurus*, 85
—*pardalis*, 85–86
Acontias, 203–205
—*meleagris*, 203–204
—*percivali occidentalis*, 204
—*plumbeus*, 204–205
Adriatic Wall Lizard, 152–153
African Slender-Snouted Crocodile, 269
African Whip Lizard, 53–54
Algerian Psammodromus, 163–164
Algyroides, 86–88
—*fitzingeri*, 86
—*marchi*, 86–87
—*moreoticus*, 87–88
—*nigropunctatus*, 88
Alligator, 256–259
—*mississippiensis*, 256–257
—*sinensis*, 257–259
Alligatoridae, 253, 255–259
Alligator Lizards, 37–41
Alligators, 253, 255–259
Ameiva, 229–230
—*ameiva*, 229–230
Ameivas, 229–230
American Alligator, 256–257
American Crocodile, 268–269
Amphisbaenidae, 243–248
Angolosaurus, 46
—*skoggi*, 46
Anguidae, 37
Anguinae, 43–45

Anguis, 44–45
—*fragilis*, 44–45
Armenian Sand Lizard, 102
Armoured Sungazer, 67–69
Atlantic Canary Island Lizard, 92–93
Australian Freshwater Crocodile, 270
Azerbaijan Lizard, 125–126

Balearic Wall Lizard, 151–152
Beaded Lizards, 3–8
Bengal Monitor, 12–13
Berber Lizard, 122–123
Bipedidea, 243
Black and White Tegu, 238–241
Black Caiman, 263–264
Black-Eared False Sungazer, 80–82
Black-Headed Monitor Lizard, 36
Blanus, 244–247
—*cinereus*, 245–246
—*strauchi*, 246–247
Blue-Black Plated Sand Lizard, 47–48
Blue-Tailed Skink, 216–217
Blue-Tongued Skinks, 181–187
Blunt-Headed Lizard, 141
Bocage's Wall Lizard, 147–148
Bosk's Fringe-Fingered Lizard, 84
Bridged Lizards, 249
Broad-Headed Skink, 194–195
Broad-Snouted Caiman, 262–263
Brown Plated Lizard, 49–51
Brown Rock Skink, 177
Bush Crocodile, 223–224

Caiman, 259–264
Caiman, 259–264

Caiman—continued
—*crocodilus*, 259-262
—*latirostris*, 262-263
Caiman Lizard, 235-236
Callopistes, 230-232
—*maculatus*, 231-232
Canary Island Lizards, 91-97
Canary Island Skink, 192-193
Cape Mabuya, 211-213
Cape Scelotes, 222-223
Cape Snake Lizard, 66-67
Caspian Green Lizard, 129-130
Caucasian Lizard, 105-106
Caucasian Racerunner, 91
Chalcides, 188-193
—*bedriagai*, 188
—*chalcides*, 189
—*ocellatus*, 189-190
—*polylepis*, 191
—*sexlineatus*, 191-192
—*viridanus*, 192-193
Chamaesaura, 66-67
—*anguina*, 66-67
Checkerboard Worm Lizard, 248
Chihuahuan Spotted Whiptail 232-233
Chilean Tegu, 231-232
Chinese Alligator, 257-259
Chinese Crocodile Lizard, 1-3
Clark's Caucasian Lizard, 106-107
Club-Footed Monitor, 21-22
Cnemidophorus, 232-235
—*exsanguis*, 232-233
—*gularis*, 233
—*inornatus*, 233-234
—*sexlineatus*, 234
—*tigris*, 234-235
Common Blue-Tongued Skink, 184-186
Common Wall Lizard, 153-154
Cophoscincopus, 205-206
—*durus*, 205-207
Cordylidae, 66-81
Cordylosaurus, 47-48
—*subtesselatus*, 47-48
Cordylus, 67-77
—*cataphractus*, 67-70
—*cordylus*, 70-71

—*giganteus*, 71-72
—*jonesii*, 73
—*jordani*, 73-74
—*macropholis*, 74-75
—*polyzonus*, 75
—*rhodesianus*, 75-76
—*warreni*, 76-78
Corucia, 170-171
—*zebrata*, 170-171
Cricosaura, 224-225
—*typica*, 224-225
Crimean Wall Lizard, 159-160
Croatian Lizard, 110-111
Crocodile Caiman, 259-262
Crocodile Lizards, 1-3
Crocodiles, 253-255
Crocodylia, 268-278
Crocodylidae, 268-278
Crocodylus, 268-278
—*acutus*, 268-269
—*cataphractus*, 269
—*intermedius*, 269-270
—*johnstoni*, 270
—*mindorensis*, 270-272
—*moreletti*, 272
—*niloticus*, 272-273
—*novaeguineae*, 273-274
—*palustris*, 274-275
—*porosus*, 276-277
—*rhombifer*, 277-278
—*siamensis*, 278
Cuban Crocodile, 277-278
Cuban Night Lizard, 224-225
Cunningham's Skink, 172-173
Cylinder Skinks, 188-193

Danford's Lizard, 107
Dark Lanceolate Skink, 203-204
Dasia, 207-210
—*olivacea*, 208-210
—*smaragdina*, 207-208
Derjugini's Lizard, 107-110
Desert Lizards, 22-23, 137-139
Desert Monitor Lizard, 22-23
Desert Night Lizards, 227-228

Index

Desert Skink, 175–176
Dracaena, 235–236
—*guianensis*, 235–236
Dumeril's Monitor Lizard, 14–15
Dunes Lizard, 136–137
Dwarf Caiman, 265–266
Dwarf Crocodile, 279–280
Dwarf Sungazer, 70–71
Dwarf Tegu, 237–238

Eastern Emerald Lizard, 114–116
Eastern Glass Lizard, 43
Eastern Golden Skink, 211
Egernia, 171–180
—*cunninghami*, 172–173
—*depressa*, 173–174
—*hosmeri*, 174–175
—*inornata*, 175–176
—*major*, 176–177
—*saxatilis*, 177
—*stokesi*, 177–178
—*striolata*, 178–180
—*whiti*, 180–181
Eisentraut's Canary Island Lizard, 94–95
Emerald Lizard, 133–134
Emerald Monitor, 27–28
Emerald Racerunner, 167–168
Emerald Tree Skink, 207–208
Eremias, 89–91
—*pleskei*, 89–90
—*strauchi*, 90–91
—*velox caucasia*, 91
Eumeces, 193–198
—*fasciatus*, 193–194
—*inexpectatus*, 194
—*laticeps*, 194–195
—*obsoletus*, 195–196
—*schneideri*, 196–198
European Common Lizard, 134–136
European Fringe-Fingered Lizard, 85
European Green Lizard, 133–134
European Sand Lizard, 98–100
European Snake-Eyed Skink, 202–203

False Cape Sungazer, 79–80
False Gharial, 280–281
False Sungazers, 79–82
Fernand's Magnificent Skink, 218–220
Fernand's Slender Skink, 218–220
Five-Lined Mabuya, 216–217
Five-Lined Skink, 193–194
Flat Lizards, 78–79
Flecked Mabuya, 213–214
Flecked and Striped Racerunner, 145–146
Florida Worm Lizard, 247
Four-Striped Girdled Lizard, 64
Four-Striped Mabuya, 214–216
Four-Toed Lizard, 54
Freshwater Crocodile 273–274
Fringe-Fingered Lizards, 83–86

Gallotia, 91–97
—*atlantica*, 92–93
—*g. galloti*, 93
—*g. caesaris*, 94
—*g. eisentrauti*, 94–95
—*g. gomerae*, 95
—*g. palmae*, 96
—*stehlini*, 96–97
Ganges Gharial, 281–283
Gavialis gangeticus, 281–283
Gerrhontinae, 38–41
Gerrhonotus, 38–41
—*coeruleus*, 38–39
—*kingi*, 39–40
—*multicarinatus*, 40–41
Gerrhosaurus, 48–53
—*flavigularis*, 48–49
—*major*, 49–51
—*nigrolineatus*, 51
—*validus*, 51–53
Gharials, 253
Giant Ameiva, 229–230
Giant Canary Island Lizard, 96–97
Giant Emerald Lizard, 130–132
Giant Lanceolate Skink, 204–205
Giant Skinks, 170–187
Giant Sungazer, 71–72
Gidgee Skink, 177–178
Gila Monster, 5–8
Girdled Lizards, 57–66
Glass Lizards, 38, 41–43

Gold Mabuya, 211
Gomera Canary Island Lizard, 95
Gould's Monitor Lizard, 18-19
Gran Canaria Skink, 191-192
Granite Lizard, 225-226
Granite Night Lizard, 225-227
Great Plains Skink, 195-196
Greek Algyroides, 88
Greek Keeled Lizard, 88
Green Madagascar Girdled Lizard, 57
Green Rock Lizard, 110
Guyana Caiman Lizard 235-236

Helmeted Skink, 223
Heloderma, 3-8
—*horridum*, 3-5
—*suspectum*, 5-8
Helodermatidae, 3-8
Hierro Canary Island Lizard, 94
Hosmer's Skink, 174-175

Iberian Emerald Lizard, 128-129
Iberian Mountain Lizard, 116-117
Iberian Skink, 188
Iberian Wall Lizard, 149-151
Ibiza Wall Lizard, 156-157
Ichnotropis, 97-98
—*capensis*, 97-98
Island Night Lizard, 226-227

Japanese Raceruners, 165-168
Jones' Dwarf Sungazer, 73

Karsten's Girdled Lizard, 59-61
Keeled Lizard, 55
Keeled Lizards, 86-88
Keeled Skink, 213
Keel-Tailed Lizard, 126-127
Khorat Skink, 220-221
Knob-Scaled Lizard, 1-3
Konx's Burrowing Lizard, 139-140

Lace Monitor, 36-37
Lacerta, 98
—*agilis*, 98-100
—*"armeniaca,"* 102
—*bedriagae*, 102-103
—*brevicaudata*, 100-102
—*cappadocica*, 103-105
—*caucasica*, 105-106
—*clarkorum*, 106-107
—*danfordi*, 107
—*derjugini*, 107-110
—*graeca*, 110
—*horvathi*, 110-111
—*jayakari*, 111-112
—*laevis*, 112-113
—*lepida*, 113-114
—*media*, 114-116
—*monticola*, 116-117
—*mosorensis*, 117
—*oertzeni*, 117-118
—*oxycephala*, 118-120
—*pamphylica*, 120-121
—*parva*, 121
—*parvula*, 121-122
—*pater*, 122-123
—*praticola*, 123-125
—*raddei*, 125-126
—*rudis*, 126-127
—*saxicola*, 127-128
—*schreiberi*, 128-129
—*strigata*, 129-130
—*trilineata*, 130-132
—*"unisexualis,"* 132
—*"uzelli,"* 132-133
—*viridis*, 133-134
—*vivipara*, 134-136
Lacertidae, 82-136
Lacertids, 82-136
Lamprolepis smaragdina, 207-208
Lanceolate Skinks, 203-205
Land Mullet, 176-177
Large Girdled Lizard, 63
Large-Scaled Sungazer, 74-75
Little Striped Whiptail, 233-234
Long-Tailed Rock Monitor, 21-22
Lygosominae, 202-203

Mabuya, 210-218
—*aurata septemtaeniata*, 211
—*capensis*, 211-213

Index

—*carinata*, 213
—*macularia*, 213-214
—*multifasciata*, 214-216
—*quinquetaeniata*, 216-217
—*vittata*, 217-218
Mabuyas, 210-218
Madagascan Keeled Lizard, 55
Madeira Wall Lizard, 148
Madrean Alligator Lizard, 39-40
Magnificent Skinks, 218-219
Mangrove Monitor, 23
Meadow Lizard, 123-125
Melanosuchus niger, 263-264
Meroles, 136-141
—*anchietae*, 136-137
—*cunneirostris*, 137-139
—*knoxii*, 139-140
—*suborbitalis*, 140-141
Mertens' Water Monitor, 25-26
Mochlus, 218-220
—*fernandi*, 218-220
—*sundevalli*, 220
Monitor Lizards, 9-37
Montane Dwarf Tegu, 237-238
Moorish Worm Lizard, 245-246
Morelet's Crocodile, 272
Moroccan Skink, 191
Mosor Lizard, 117
Mugger Crocodile, 274-275
Multi-Banded Sungazer, 75

Namaqua Racerunner, 146
New Guinea Helmeted Skink, 223-224
Night Lizards, 224-228
Night Skinks, 199-200
Nile Crocodile, 272-273
Nile Monitor, 26-27
Northern Alligator Lizard, 38-39
Nucras, 141
—*intertexta*, 141-142

Oak Skink, 181-182
Ocellated Lizard, 113-114
Ocellated Skink, 189-190
Olive Tree Skink, 208-209
Oman Lizard, 111-112

Ophiomorus, 198-199
—*punctatissimus*, 198-199
Ophisaurus, 41-43
—*apodus*, 41-43
—*ventralis*, 43
Ophisops, 142-144
—*elegans*, 142-143
—*occidentalis*, 143-144
Orinoco Crocodile, 269-270
Ornate Girdled Lizard, 63-64
Osteolaemus, 279-280
—*tetraspis*, 279-280

Paleosuchus, 264-267
—*palpebrosus*, 265-266
—*trigonatus*, 266-267
Palma Canary Island Lizard, 96
Pamphylic Emerald Lizard, 120-121
Pedioplanus, 144-146
—*breviceps*, 144-145
—*lineoocellata*, 145-146
—*namaquensis*, 146
Peleponnese Wall Lizard, 154-156
Perentie, 19-20
Peters' Plated Lizard, 55
Philippine Crocodile, 270-272
Pholidobulus, 237-238
—*montium*, 237-238
Pine Cone Skink, 186-187
Pink-Tongued Skink, 182-183
Plated Sand Lizards, 46
Platysaurus, 78-79
—*guttatus*, 78-79
Podarcis, 146-161
—*bocagai*, 147-148
—*dugesii*, 148
—*erhardii*, 149
—*hispanica*, 149-151
—*lilfordi*, 151-152
—*melisellensis*, 152-153
—*muralis*, 153-154
—*peloponnesiaca*, 154-156
—*perspicillata*, 156
—*pityusensis*, 156-157
—*sicula*, 157-159
—*taurica*, 159-160

Podarcis—continued
—*tiliguerta*, 160
—*wagleriana*, 160–162
Prehensile Tailed Skink, 170–171
Psammodromus, 162–165
—*algirus*, 163–164
—*hispanicus*, 164–165
Pseudocordylus, 79–82
—*capensis*, 79–80
—*melanotus*, 80–82
—*microlepidotus*, 82
Pygmy Algyroides, 86
Pygmy Mulga Monitor, 20–21
Pygmy Spiny-Tail Skink, 173–174

Racerunners, 89–91
Red-Footed Girdled Lizard, 65
Red-Throated Running Tegu, 233
Rhineura floridana, 247
Rhineuridae, 243
Rhodesian Sungazer, 75–76
Rhynchocephalia, 249–252
Ridgetail Monitor, 10–11
Ringed Worm Lizards, 244–245
Riopa, 220–221
—*koratense*, 220–221
Riopa Skinks, 220–221
Rock Lizard, 127–128
Rough-Necked Monitor, 28–29
Rough-Scaled Cape Lizard, 97–98
Ruin Lizard, 157–159

Saltwater Crocodile, 276–277
Sandfish, 200–201
Sand Lizards, 136–141
Sandrunner, 162–163
Sand Skinks, 200–201
Sardinian Algyroides, 86
Sardinian Keeled Lizard, 86
Savannah Monitor Lizard, 16–17
Scelotes, 221–223
—*bipes*, 221–222
—*capensis*, 222–223
Scelotes Skinks, 221–223
Scheltopusik, 41–43
Schreiber's Green Lizard, 128–129

Scincidae, 169–244
Scincopus, 199–200
—*fasciatus*, 199–200
Scincus scincus, 200–201
Sharp-Snouted Lizard, 118–120
Shinisaurus, 1
—*crocodilurus*, 1–3
Short-Headed Racerunner, 144–145
Short-Tailed Monitor, 13–14
Short-Tailed Sand Lizard, 100–102
Siamese Crocodile, 278
Sicilian Wall Lizard, 160–162
Six-Lined Racerunner, 165–166, 234
Six-Lined Skink, 191–192
Skinks, 169–244
Slender Skink, 220
Slender Skinks, 202–207
Slow Worm, 43–45
Small Flat Lizard, 78–79
Small-Scaled False Sungazer, 82
Smith's Monitor Lizard, 23–25
Smooth-Headed Caiman, 266–267
Smooth Lizards, 169–244
Smooth-Necked Mountain Lizard, 169
Snake-Eyed Lizard, 142–144
Snake-Eyed Skink, 202–203
Snake Skinks, 198–199
South African Sungazer, 71, 72
Southeastern Five-Lined Skink, 194
Southern Alligator Lizard, 40–41
Southern Mountain Lizard, 168
Spanish Algyroides, 86–87
Spanish Keeled Lizard, 86–87
Spanish Psammodromus, 164–165
Speckled Snake Skink, 191–192
Spectacled Caiman, 259–262
Spectacled Wall Lizard, 156
Spencer's Monitor, 31–32
Spenops sepsoides, 202
Sphenodon punctatus, 249
Spiny Skink, 171
Spotted Skink, 196–198
Spotted Tree Monitor, 34–36
Storr's Dwarf Monitor Lizard, 32–33
Storr's Monitor Lizard, 32–33
Strauch's Racerunner, 90–91

Index 307

Striped Canary Island Skink, 191–192
Striped Green Lizard, 129–130
Striped Mabuya, 217–218
Striped Plated Lizard, 51
Stripe-Tailed Monitor, 14
Sudan Plated Lizard, 49–51
Sunda Gharial, 280–281
Sundevall Magnificent Skink, 220
Sungazers, 66–78
Syrian Lizard, 112–113

Takydromus, 165–168
—*sexlineatus*, 165–167
—*smaragdinus*, 167–168
Tegus, 238–241
Teiidae, 229–241
Tetradactylus, 53–55
—*africanus*, 53–54
—*breyeri*, 54
—*seps*, 54
—*tetradactylus*, 54–55
Texas Spotted Whiptail, 233
Three-Striped Girdled Lizard, 65–66
Tiliqua, 181–187
—*casuarinae*, 181–182
—*gerrardi*, 182–183
—*multifasciata*, 184
—*rugosa*, 186–187
—*scincoides*, 184–186
Tiliquinae, 170–187
Timor Monitor Lizard, 33–34
Tomistoma, 280–281
—*schlegelii*, 280–281
Tracheloptychus, 55–57
—*madagascariensis*, 55
—*petersi*, 55–57
Transcaucasian Racerunner, 89–90
Tree Skink, 178–180, 207–208
Tribolonotus, 223–224
—*novaeguineae*, 223–224
Trogonophidae, 247–248
Trogonophis wiegmanni, 248
Tropidosaura, 168–169
—*gularis*, 168–169
—*montana*, 169

True Crocodiles, 268–278
True Sharp-Nosed Skinks, 225
True Worm Lizards, 244–248
Tuataras 249–252
Tupinambis, 238–241
—*teguixin*, 238–241
Turkish Worm Lizard, 246–247
Two-Toed Scelotes Skink, 221–222
Tyrrhenian Sand Lizard, 102–103
Tyrrhenian Wall Lizard, 160

Varanus, 9–37
—*acanthurus*, 10–11
—*bengalensis*, 12–13
—*brevicauda*, 13–14
—*caudolineatus*, 14
—*dumerilii*, 14–15
—*exanthematicus*, 16–17
—*flavescens*, 17–18
—*flavirufus*, 18–19
—*giganteus*, 19–20
—*gilleni*, 20–21
—*glebopalma*, 21–22
—*gouldii*, 18
—*griseus*, 22–23
—*indicus*, 23
—*karlschmidtii*, 23–25
—*mertensi*, 25–26
—*niloticus*, 26–27
—*prasinus*, 27–28
—*rudicollis*, 28–29
—*salvator*, 29–30
—*similis*, 34–36
—*spenceri*, 31–32
—*storri*, 32–33
—*timorensis*, 33–34
—*tristis*, 36
—*varius*, 36–37
Virgin Lizard, 132
Viviparous Lizard, 134–136

Wall Lizards, 146–161
Warren's Sungazer, 76–78
Water Monitor, 29–30
Water Skink, 205–206
Wedge-Headed Skinks, 201–202

Western Girdled Lizard, 61
Western Lanceolate Skink, 204
Western Snake-Eyed Lizard, 143–144
Western Whiptail, 234–235
Whip Lizards, 53
Whiptail Lizards, 232–235
White's Skink, 180–181
Worm Lizards, 243–248

Xantusia, 225–228
—*henshawi*, 225–226
—*riversiana*, 226–227
—*vigilis*, 227–228
Xantusidae, 224–228
Xenosauridae, 1

Yucca Night Lizard, 227–228
Yellow Monitor Lizard, 17–18
Yellow Plated Lizard, 48–49

Zonosaurus, 57–66
—*aeneus*, 57
—*haraldmeieri*, 57–59
—*karsteni*, 59–61
—*laticaudatus*, 61
—*m. madascariensis*, 61–63
—*maximus*, 62–63
—*ornatus*, 63–64
—*quadrilineatus*, 64–65
—*rufipes*, 65
—*trilineatus*, 65–66